SELF-SERVICE ANALYTICS SIMPLIFIED

How to Plan and Implement

ARSHAD KHAN

Self-Service Analytics Simplified: How to Plan and Implement
Arshad Khan
San Jose, California, USA

ISBN-13: 978-0-9660863-5-5

Typeset by Amnet Systems
Cover design by Amnet Systems

Published by Khan Consulting and Publishing LLC

TABLE OF CONTENTS

ABOUT THE AUTHOR

Arshad Khan is a versatile business intelligence professional with over 30 years of experience. He has consulted for Accenture, PwC, Deloitte, Navy, Chevron, Bose Corporation, Daimler Chrysler, Genentech, PepsiCo, Hitachi-America, and many other blue-chip companies. Mr. Khan, who is currently working as a self-service analytics consultant for UCSF, has led numerous analytics projects for many organizations.

Mr. Khan has sixteen books to his credit, including seven books on BI/ analytics. He has taught as an adjunct professor at seven universities including Golden Gate University and the University of California (San Diego, Berkeley, and Santa Cruz). Mr. Khan has a graduate degree in engineering and an MBA.

PREFACE

Self-Service Analytics Simplified: How to Plan and Implement will introduce you to self-service analytics (SSA), which aims to make business users more productive and less dependent on IT for their reporting and analytics needs. This book, which teaches how to plan and implement an SSA project, will appeal to a broad range of users including senior executives, business and IT managers, project managers, data analysts, business analysts, developers, power and casual users, as well as IT professionals.

A wide range of topics are covered in *Self-Service Analytics Simplified: How to Plan and Implement*. These include an introduction to self-service analytics, relationship with business intelligence, the need, benefits and capabilities of a data analytics platform, as well as the benefits for the organization, business users, and IT.

Self-Service Analytics Simplified shows how to assess if an organization is ready for an SSA platform and also how to plan and implement an SSA project. It describes in detail the different types of users expected to use the SSA platform and how they can be mapped to the architecture. Since data is the lifeblood of an SSA platform, various data-related topics are also covered including metadata, data pipelining, and governed data flow.

Self-Service Analytics Simplified: How to Plan and Implement describes the SSA architecture in detail, as well as its components and tools. It also covers data governance, security, training, data and user onboarding, and barriers to adoption, as well as challenges, common mistakes, best practices, lessons, and tips.

Hope this book gets you started on the road to analytics independence. Happy reading!

Arshad Khan

Chapter 1
UNDERSTANDING ANALYTICS

What is business intelligence?

Business intelligence (BI) is a decision-support system whose aim is to help make business decisions, strategic as well as operational/tactical. It uses an assortment of resources and techniques for gathering, transforming, storing, and analyzing data. BI is an umbrella term that includes the applications, infrastructure, tools, and best practices that enable access to and analysis of information for optimizing decisions and performance[1].

What is analytics?

Analytics as the discovery, interpretation, and communication of meaningful patterns in data and the process of applying those patterns towards effective decision-making. In other words, analytics can be understood as the connective tissue between data and effective decision-making within an organization. Analytics relies on the simultaneous application of statistics, computer programming, and operations research to quantify performance[2].

Business analytics and business intelligence are different, even though they have some overlap. While both use metrics, BI tends to evaluate past

1 Khan, A (2012), *Business Intelligence and Data Warehousing Simplified*, Mercury Learning

2 https://en.wikipedia.org/wiki/Analytics, Wikipedia

performance based on some specific and consistent metrics, which are then used for decision-making. Business analytics uses a broader range of metrics which, rather than focus on what has already happened, focuses more on what will happen. Analytics is also more complex because it uses more complex processes and algorithms.

Types of analytics

Descriptive Analytics

Descriptive analytics is the most common and simplest type of analytics, which accounts for about 80% of business analytics. Its objective is to summarize what has happened so that a user can interpret the results, which can enable insights to be derived from it. In this type of analytics, data from past periods is summarized and converted into a form that can be easily understood and interpreted by a user. An example is a retailer who, based on the available data, wants to determine the most profitable mix of electronics for the Christmas season.

Diagnostic Analytics

Diagnostic analytics is a more advanced type of analytics, which logically follows descriptive analytics. Instead of focusing on what has happened, it focuses on why something happened. It tries to answer questions like why did something happen? What caused it? For example, using diagnostic analytics, a company's increased sales in Asia could be attributed to the marketing campaign conducted in the first quarter of the year, or a competitor leaving the market. A number of techniques can be used for diagnostic analytics, such as correlation, data mining, and data discovery.

Predictive Analytics

This is an advanced type of analytics, which is used to predict the future, based on current events. Using advanced techniques, it first determines patterns or relationships in large data volumes and, subsequently, extrapolates them into the future to predict events and behaviors. As an example, predictive analytics can be used to predict the likelihood of a diabetic patient having a heart attack in the next ten years.

Predictive analytics uses advanced tools and techniques such as data mining, statistics, modeling, machine learning, and artificial intelligence, as well as clustering, decision trees and neural networks.

Prescriptive Analytics

Prescriptive analytics uses advanced technologies to determine the best solution or course of action, from among many choices. It anticipates what, when, and why something will happen. Prescriptive analytics can be used to benefit from a future opportunity or avoid a future risk. Using prescriptive analytics, an analyst can decide what to do with a trend or issue that was forecasted by predictive analytics. Prescriptive analysis uses a combination of tools and techniques, which include its own logic and algorithms, machine learning, as well as modelling procedures.

Analytics roadmap

The data strategy that an organization develops needs to be converted into an actionable plan, which an analytics roadmap provides. The plan enables the implementation of the key initiatives of the data strategy. The roadmap ensures that all strategic initiatives are evaluated from different perspectives, including constraints and value. It should include:

- Tasks, activities and milestones required for implementing the data strategy
- Dependencies
- Resources and skills
- Timeline (schedule)
- Implementation phases, as well as their details

The analytics objective must be in sync with the organization's strategic plans. This will help determine the outcomes, such as implementing a single reporting and analytics infrastructure or providing a single version of the truth. After the program outcomes have been decided and agreed upon, they are prioritized based on a number of factors, such as their impact on the business, technical complexity, cost, etc.

Building an analytics roadmap

The building an analytics roadmap requires the following:

- Comprehensive review: Covers various business aspects, identifies the risks, and outlines mitigation plans
- Is tied to business goals: Identifies the business goals, ties the analytics to the business goals and achieves consensus amongst business owners
- Clear understanding of existing state: Outlines the current state, identifies missed opportunities, and uncovers existing challenges
- Prioritized implementation phases: Identifies required phases, prioritizes according to business importance, and provides estimates

Analytics roadmap project plan

The project plan for an analytics roadmap can be split into four stages: Business, analytical, capability and technology[3]. In the business stage, the tasks performed include understanding the strategies and objectives, conducting business process and maturity assessment, identifying pain points and priorities, shortlisting of focus areas, as well as estimating returns.

In the analytical stage, the key tasks performed include translating objectives into analytical requirements, creating a high-level solution design, developing business cases, estimating investments and returns, as well as creating the final shortlist. In the capability stage, the key tasks include developing the detailed execution roadmap, developing the skills and expertise strategy, identifying the technical tools, as well as performing readiness analysis.

3 https://www.sas.com/content/dam/SAS/en_my/doc/other2/building-analytical-roadmap.pdf

In the technology stage, the key tasks include conducting technology maturity assessment, assessing the technology gap, documenting the technology requirements, developing the vendor and cloud strategies, as well as determining the technology evolution roadmap.

Chapter 2

BUSINESS INTELLIGENCE AND SSA

Status of traditional BI

After a period of tremendous growth, the demand for traditional BI is starting to slow. However, it continues to meet the reporting needs of routine day-to-day business operations, which cover functional, operational, financial, regulatory, and compliance requirements. Traditional BI primarily answers two questions:

- What has happened (historical)
- What is happening now (operational)

While the best days of BI are over, it still provides tremendous value and will continue to benefit organizations requiring centralized reporting. Traditionally, BI has been based on data stored in data warehouses and data marts. However, the methods, architectures, and software solutions that have been considered best practices in data warehousing are now insufficient to meet the changing needs of many organizations. Therefore, they have started losing favor and newer technologies, like Hadoop, are becoming serious competitors.

Inadequacy of traditional BI environment

The objective of traditional BI was to empower business users by providing information that could not be obtained from operational systems easily or quickly. It relied on cleansed and transformed data, stored in centralized data stores like data warehouses and data marts, which could be accessed for developing reports and performing limited ad-hoc queries.

This environment was very rigid and any changes needed, post-deployment, could not be implemented easily or quickly. Users who needed new reports or wanted access to new data sources had to initiate change requests with the Information Technology (IT) department, which was responsible for the BI platform. Due to limited IT resources, such requests had to be put into a queue and, typically, took weeks or months to implement.

This problem has been compounding due to the exponential rise in the amount of data being generated and required for monitoring business performance. In a dynamic business environment, such delays are unacceptable. Therefore, both business and IT have been seeking ways to solve this serious problem that has frustrated users.

Self-service analytics

Traditional business intelligence tasks like report and dashboard development were typically performed by IT staff. In self-service analytics, these tasks are performed by the business users, using easy-to-use BI tools, with minimal or no IT involvement. In self-service analytics (SSA), which is a form of business intelligence, IT sets up the infrastructure and is responsible for the application software, databases, data sources, security, user management, performance monitoring, and other technical tasks. They also develop complex reports and dashboards, when needed. This de-centralized model provides business users self-sufficiency, flexibility and access to corporate data, while making them responsible for their own reports and dashboards.

Self-service analytics has many facets and each implementation has its own flavor, depending on the skill level of the users, enterprise and user analytics requirements, as well as desired controls pertaining to governance and security.

Defining self-service analytics

According to Gartner, self-service analytics is a form of business intelligence (BI) in which line-of-business professionals are enabled and encouraged to perform queries and generate reports on their own, with nominal

IT support. Self-service analytics is often characterized by simple-to-use BI tools with basic analytics capabilities and an underlying data model that has been simplified or scaled down for ease of understanding and straightforward data access[4].

Other SSA definitions include:

- Self-service analytics or self-service business intelligence refers to tools used to connect and analyze data, which are operated primarily by business departments in the organization – rather than IT professionals or dedicated data analysts
- Self-service analytics is an application of business intelligence in which business professionals are encouraged to extract data, generate reports, and run queries to enable professionals to extract value out of the data with no IT support, using various tools and software
- Self-service analytics/BI tasks are those that business users carry out themselves instead of passing them on to IT for fulfillment

Comparing traditional BI and self-service analytics

Traditional business intelligence is associated with legacy vendors like Oracle (OBIEE), SAP (Business Objects), IBM (Cognos) and others. Their BI tools provided robust enterprise querying and reporting capabilities. However, they did not have self-service capabilities that could be used by most users.

The key differences between traditional business intelligence and self-service analytics are highlighted in the next two sections.

Traditional Business Intelligence

Traditional BI has the following characteristics:

- Is more expensive
- Takes longer to implement

4 https://www.gartner.com/it-glossary/self-service-analytics

- Is more complex and difficult to implement
- Very data warehouse-centric
- Works on larger data volumes
- SQL is an important tool used for developing queries
- End-users are highly dependent on IT or data analysts for their reporting and analysis needs

Self-Service Analytics

Self-service analytics has the following characteristics:

- Low entry cost enables data to be analyzed quickly, even by those who are not tech savvy
- Not data-warehouse centric
- Works on smaller datasets
- Most users can perform their work using front-end tools, without ever needing to write SQL
- Provides more flexibility
- Users can perform most of their analyses, without requiring help from IT or data analysts
- Data does not need to be moved into a data warehouse before it can be analyzed
- May use tools that are not as robust or have feature limitations
- Provides access to IT-curated data sources, as well as non-traditional data sources

SSA's adjunct role

Self-service analytics does not replace traditional business intelligence. It is not a replacement for development that occurs under the traditional business intelligence umbrella. It enables business users to develop their own analytics, without dumping everything onto IT, which permits them to move faster. Traditional business intelligence, in the new environment, becomes part of the hybrid approach of centralized and decentralized models.

Self-service analytics versus self-service BI

The two terms, self-service analytics and self-service business intelligence, are usually used interchangeably. However, business intelligence is more closely associated with obtaining answers to what is happening in the business now or what happened in the past. On the other hand, analytics is more focused on what is happening now and what will likely happen in the future.

Implementing SSA

Self-service analytics must be implemented as part of a data strategy, which is an initial high-level view of an organization's objectives for implementing analytics. It provides an understanding of how data can be leveraged by the organization. It leads to an understanding of the potential opportunities and benefits, identification of the available data and its quality, data to be migrated, as well as integration. Other topics covered by the strategy include ownership, roles, responsibilities, security, governance, and organization structure.

Developing the data strategy enables the organization to understand what it wants to achieve by implementing analytics. It clarifies the high-level objectives which includes a better understanding of the business, determine what is happening now and why, as well as predicting what will happen in the future. These will help the organization optimize how the business is run in order to decrease costs, increase revenue, and improve profitability. To meet these objectives, an analytics roadmap is required.

Chapter 3

UNDERSTANDING SELF-SERVICE ANALYTICS

Objective

The objective of SSA is to make organizations data-driven, decentralize business intelligence, make business users independent and self-sufficient, enable them to perform their own reporting and analysis, make data available to the masses so that it can be easily and quickly analyzed and, also, remove IT from being a bottleneck for reporting and analysis.

Drivers

In the Internet age, there has been a push for self-service which is exemplified by booking your own travel, online banking, and online shopping. In the same way, business intelligence has been impacted by the desire of users to avoid being dependent on IT for their reporting and analytics needs. Simultaneously, an explosion in the generation of data has forced businesses to develop a data-driven culture, which has led to the SSA concept becoming very popular.

The following are other drivers for the growing popularity of self-service analytics:

- Desire for agility and flexibility that traditional BI could not provide
- Availability of powerful and easy to use front-end tools, which make users more productive and less dependent on IT

- Desire for new insights without assistance
- Desire to leverage untapped enterprise data
- Requirement to add new data sources quickly
- Provide access to data anytime and anywhere
- Reduce the IT report and dashboard development backlog and burden
- Reduce time between identifying a need to the rollout of a solution
- Perform analysis without coding or using SQL

Monetization driver

The legacy BI delivery best practices for reporting and analysis were based on the data warehousing architecture. They were popular in the past few decades, but are now inadequate to meet the needs of data-driven organizations. Data is growing exponentially and will continue to do so. For most enterprises, monetizing data is very important as it can help grow revenues. Organizations that can use their data to increase revenue, decrease costs, improve profits, and improve their competitive advantage will be winners. Achieving these goals can be easier for an agile organization using self-service analytics, which can make employees more productive and help them make decisions quickly.

Value provided by SSA

A self-service analytics platform meets the needs of business users who need data discovery, interactive reporting, and dashboards. It also provides users the ability to prepare data, which typically has been the domain of IT. At the same time, it provides security and governance that IT is responsible for. The key areas where self-service analytics provides value include:

- Ease of use: Users can perform data and analytics tasks, without having to write SQL or learn programming languages
- Agility: Enables users to quickly gain insight as they are self-sufficient, without having to wait for IT to deliver what they want

- Flexibility: Users can tap a variety of data sources of different types, structures, and size
- Scalability: Analysis can be performed across departments, geographic areas, and hierarchies, within governance rules and standards

Evolution of SSA

Business intelligence software was very difficult to use and restricted to a small number of users who were typically part of IT. It required technical skills that developers and data scientists had—not ordinary business users. Business users could not even develop simple reports and dashboards that they needed. Therefore, most of the demand was limited to senior executives and managers—not the users lower in the corporate hierarchy. Also, since the available tools were not easy to use and had very limited data visualization capabilities, they could only be used for developing simple charts and graphs.

When the demand for reporting and analytics increased, the IT backlog started to increase. This, combined with the ordinary users' demand to be empowered, led to the evolution of the self-service analytics platform. In a relatively very short period, it has become very popular. Its availability has changed the way organizations perform analyses and decision-making. It is estimated that self-service analytics users will soon produce more analyses than data scientists.

Business and IT partnership

Eliminating the role of IT is not the goal of a self-service analytics program. Instead, it aims to increase collaboration between the business and IT, which has not been very productive in the past. SSA enables ad-hoc and simpler reports as well as analyses to be offloaded to the business users, which reduces the IT workload. Instead of providing requirements to IT, which then puts them in a queue to be developed after a lengthy wait, business users can perform prototyping and development on their own.

Since IT understands data very well, better than the business, it provides the curated data in the required format and frequency to the business users. It also rolls out into production the objects that the business users develop, such as reports and dashboards, after ensuring that they are production ready. Validation performed through user acceptance testing (UAT) and by IT ensures that any objects being developed or modified do not create quality or technical performance issues.

The IT and business partnership is a win-win for both sides. SSA initiates the business requirements, which helps produce the output quickly and in an optimal way. When technical complexity develops, IT can bring in their expertise and help the users.

Misunderstanding SSA

Self-service analytics is often misunderstood. The following are some common misconceptions:

- It is a niche tool, which is not correct because SSA is a full-fledged platform that empowers all types of users
- It is a data visualization tool, which is not correct because it has far more features and capabilities
- All data is available and accessible just through a few clicks, which is not correct as data needs to be onboarded through a process
- Replaces traditional business intelligence, which is not correct as SSA complements, rather than replace traditional BI
- Eliminates the need for data warehouses, which is not correct as SSA uses data warehouses as well as other data sources
- Is dependent on combined datasets, which is not correct as SSA users can work on small data sources like Excel, as well as larger enterprise level data sources (ERP systems, data warehouses, and data marts)
- Eliminates the need for the extraction, transformation and loading (ETL) process, which is not correct as those processes are still required
- Requires SQL for effective use, which is not correct as most users may never use SQL to analyze data and produce reports

- Is shadow IT, which is not correct as SSA is a platform where there is close collaboration between IT and the business
- Does not provide governance, which is not correct as SSA supports governance

Chapter 4

DATA ANALYTICS PLATFORM

Need for a data analytics platform

Data, which is generated by every organization and individual, is a very valuable asset. The volume of data that any organization generates and captures is huge. This data is also very rich and varied. Therefore, businesses need to analyze it in order to remain competitive, understand their customers, increase revenue, cut costs, and improve their overall performance.

The availability of this data, by itself, does not provide any benefit. It has to be utilized efficiently and quickly, which is a challenge due to the disparate data sources where data is stored in silos and not easily available for access or analysis. This can only be done if there exists a platform where data can be prepared and then fed to the users, who can then utilize it for their work to get a better understanding of the organization's market, customers, and business operations.

Benefits of a self-service analytics platform

A dedicated SSA platform's key benefits are that it:

- Provides the infrastructure for pulling in raw data, transforming it, and then serving it for data visualization, reporting, and analytics

- Provides ability to access data from different types of data sources (such as databases, data marts, data warehouses, data lakes, social media, etc.)
- Enables users to access different platform components, depending on their authorized level and permissions, and then work with the data at their own skill level to produce the desired reports or analyses
- Graphically visualizes data that has been readied on dashboards or in reports
- Provides a foundation for executing well-informed business decisions

Data analytics platform capabilities

The fundamental requirement of an SSA platform are that it must:

- Be able to ingest, store, and serve data seamlessly
- Be secure
- Provide governance
- Provide an analytics tool for reporting and analysis
- Be flexible to support different user roles
- Provide collaboration capabilities

The capabilities of the SSA platform and how it has been implemented determine how easily and effectively its users can leverage the available data. With a well-designed platform, the time needed before insights can be generated will be reduced and, also, it will be cost effective for the business as well as IT. The specific capabilities depend on the architecture as well as the individual components, due to the numerous available options. For example, storage can be implemented as an on-premise database, which requires servers that may not scale easily as required by a data-driven organization. On the other hand, it can be implemented in the cloud which, besides storage flexibility, also provides high scalability.

Factors favoring SSA platform over BI platform

The traditional business intelligence platform is very structured and closed. IT manages the complete data lifecycle. Users get to work with the data only after IT provides access. The type of work that they can perform is constrained by the available data, which may not be sufficient for analysis. If a user wants to add another data source, dimension, or measure, a request will need to be initiated for IT to make the change.

In such a platform, users can only work in the target database, not in the landing zone or the staging area. Therefore, the traditional business intelligence platform is inadequate and an SSA platform is more appropriate when users want to:

- Access different parts of the platform
- Experiment in the staging area and work with raw data
- Add new dimensions and measures
- Add more data sources
- Prepare their own data
- Perform ad hoc analysis
- Develop their own reports and dashboards
- Manage governed data
- Have one version of the truth

When the data needs of users are not met, it leads to the creation of shadow IT groups, who install their own analytics tools, like Tableau and Qlik, to support their reporting and analysis needs. The results generated by these groups are not shared and each group only trusts its own data. Therefore "one version of the truth" becomes a casualty. Such a scenario is a good indicator that a single analytics platform is needed.

Another key factor that favors an SSA platform is scalability. While a traditional business intelligence platform meets many reporting needs, it is not as scalable as an SSA platform. When an enterprise requires a single location, which provides a 360-degree view for data analysis, an SSA platform is favored.

An SSA platform can extract and consolidate data from disparate systems using different storage systems (data lakes, data marts, data warehouses, cloud data warehouses, and databases), as well as files, different formats, multiple APIs, etc. After the platform makes the data available, its users can utilize it to meet their own individual reporting and analytics needs, as well as those of the enterprise. If this capability is lacking, it is also a good indicator that an SSA platform may be an appropriate solution.

Chapter 5

SSA BENEFITS FOR THE ORGANIZATION

Empowered employees

With self-service analytics, employees across the enterprise are empowered as they can extract information as needed, with limited constraints, which leads to improved and faster decision-making. The empowerment extends from the executive suite to those at the lower levels in the organization's hierarchy.

Agility

Self-service analytics provides many benefits to users who, despite sitting on a data gold mine, are limited in the information that they can access and utilize. A key objective of providing self-service is to improve the agility of users as well as the organization. Since SSA reduces their dependence on IT, business users can quickly perform routine or ad hoc analysis. The result is that the business ends up owning their analytics and derive maximum value through an environment that enables agility.

Better insight

A key objective of self-service analytics is to empower users so that they can use the available data to gain insight into the business from different perspectives. When implemented, SSA can provide most of the data required to power the business users, without requiring help from IT. For the organization, it is an important benefit as even ordinary users with limited technical skills can perform very powerful analysis which, previously, could only be performed by data analysts. They can develop their own reports and dashboards without requiring the help of developers, who are in short supply.

When an organization is provided meaningful insights, it can make data-driven, sound tradeoff decisions regarding cost, schedule, quality, etc.

More time for analysis

The most difficult part in the information cycle is to identify, extract, validate, and store data in a format that can be used by the business users. For most of the data, these tasks are performed by IT. Therefore, since the most difficult part of the cycle has already been completed, business users can access the data easily. Since they do not have to expend time on data collection, they can spend more time on analysis, which is more beneficial for the organization.

Single version of the truth

A single version of the truth refers to the single view of data that everyone in an organization agrees to and trusts. This is challenging to achieve because an enterprise has multiple types of data sources such as data warehouses, data marts, online transaction processing (OLTP) databases, Excel spreadsheets, etc. When different departments, groups, and enterprise users access this data, they will usually generate different results. The challenge is to serve this data in a way so that enterprise users trust it as the single source of the truth.

Self-service analytics provides decision-makers with data that is consistent and trusted. When such data is provided, they can focus on analysis and providing insight, rather than manipulating data so that it is in the format required for analysis. This can be game changing for organizations and users who, in the past, have spent an inordinate amount of time making sense of data stored in spreadsheets and databases with suspect data quality. When a single version of the truth is available, it provides consistency and transparency and, also, makes users more productive.

The challenge is to maintain the single version of the truth in an environment where new data sources continue to be added and existing sources are in constant update mode.

Increased productivity

Self-service analytics increases productivity for a variety of reasons, including:

- Faster decision-making
- Better quality of analysis and decision-making
- Business requirements can be less formal as ad-hoc analysis becomes more popular and, also, IT does not have to be provided formal business requirements
- Analytics initiatives can be integrated with routine day-to-day business
- Less expensive tools can usually meet the needs of the vast majority of users
- Focus on value-added analytics is beneficial for productivity
- Errors are reduced as curated data is made available to different groups and departments
- Manual execution of reports is reduced

Cost reduction

As adoption rates increase, the cost of infrastructure and systems can be spread over a larger user population and economies of scale kick in. Costs associated with bringing data into a common storage system, like a data warehouse, are reduced. In many cases, licensing costs are decreased.

When data is brought into a governed system, it becomes available to everyone in the organization. If the data/metadata is well documented, new departments do not have to repeat the process undertaken by the earlier departments. The reason is that governed data is available to all users in the organization, subject to restrictions applicable for PII (Personal

Identifiable Information), PHI (Protected Health Information) and other sensitive data. This reduces the time for onboarding and analysis, which helps reduce costs.

Repeatable workflows

It is well known that the most time-consuming steps in serving data to the users include preparing, cleansing, transforming, and blending data. By creating analytics workflows, many tasks can be automated. For example, an extract that is needed for a dashboard can be scheduled to run at a specific time or based on a trigger. Modern front-end tools enable live connections to the data sources, while extracts can be used when performance becomes an issue. These extracts can be scheduled, as needed, through workflows.

Skills-based focus on tasks

When a self-service analytics environment is available, business users perform routine and less technical tasks. These include core analytics tasks such as data exploration, data visualization, ad-hoc querying, and reporting. Therefore, data analysts and data scientists can concentrate on the tasks that they specialize in, which are more technical and complex. This provides value to the organization because employees can utilize their skills for tasks for which they are better suited.

Data democratization

In the past, business intelligence tools primarily targeted IT developers, analysts, and tech-savvy users. These users made up a small percentage of the overall user base. With the advent of self-service analytics and easy to use tools for ordinary users, enterprise data mines opened up for the large number of users who could not previously access such data. This has led to what is commonly referred to as data democratization, which implies that data is available to the vast majority of users in an enterprise.

In self-service analytics, due to data democratization, the focus is no longer on gathering, processing, and interpreting data by a small group of users. Data democratization has allowed business users, who understand the business better than IT and developers, to explore that data and produce meaningful analyses. Such users not only know which data they want, they also know how to use it. They no longer have to depend on others to provide them results and analyses, as they can access the data directly to get what they want.

In many cases, business users can connect directly to disparate data, external or internal, so long as they are authorized by IT, which is responsible for maintaining security and ensuring that governance rules are adhered to. In the right environment, this can enable more skilled users to become data scientists and, therefore, have the capability to perform more in-depth and cutting-edge analysis.

Blending and importing data

SSA provides many data benefits to the users as well as the organization, as everyone can use the same data. Self-service capabilities are supported by a common, validated, and consistent data repository that is always current and available to everyone in the enterprise. In many cases, SSA can enable the blending of data from multiple data sources, quickly and easily, using simple point and click. The data provided is of better quality and accuracy compared to what has been the norm. New or refreshed data can be pulled into analytic models, without requiring cumbersome reprocessing of the data or modification of the existing workflow.

Another data benefit is that new datasets can be brought in quickly from external sources or internally. These datasets can be exported in a number of formats and, also, written back to databases when needed to support downstream activities or analyses. different models can be tested quickly and iterated as needed. Also, reports and dashboards can be scheduled for updates, refreshes, or distribution based on data changes.

Data security

Data is more secure in an SSA environment due to controls and data governance. Access to data can be provided on a need to know basis. Authorization for access to the data can be implemented by department, group, user, hierarchy, row level, etc. This means that the data can be made as secure as needed.

Improved data quality

Data issues and poor data quality is a common issue, which can have severe consequences and lead to faulty decision-making. In a governed self-service analytics environment, data quality is superior and, therefore, can be used with confidence. With metadata management and tools, such as a data catalog, users are provided with validated data that can be used with confidence. For the business users, life is simplified because they can use definitions and calculated values that are standardized and, therefore, can be trusted and used with confidence.

Chapter 6

SSA BENEFITS FOR BUSINESS USERS

Empowerment

An SSA platform enables reporting and analytics for non-technical users. In such an environment, business users can interact with their reports, perform customization, and conduct in-depth data analysis for achieving better insight into their business. They can perform a range of operations and analysis on the data which, in the past, was not possible for them. These include filtering, drilldown, drill across, sorting, downloading data, creating crosstabs, visualizing data through a wide variety of charts and graphs, exporting to Excel or other applications, trending, forecasting, and incorporating parameters. Employing easy to use tools, these functions can be performed though simple drag and drop operations. End-users can also customize views, create alerts for abnormal conditions (exceptions), share reports and dashboards with other users, and schedule reports.

Flexibility

Just like pumping your own gas or ordering your air tickets online, self-service analytics provides flexibility to all types of users. The democratization of data provides flexibility to users who can utilize it to develop insights and address business challenges quickly and effectively, without dependence on IT or data scientists. A self-service analytics platform can enable users to get the data that they want, in the format that they desire, when they want it, where they want it, and also share it with other users.

Reporting

Reports and dashboards can be provided on different devices such as desktops, mobile devices, or the web. They can be interactive with drill-down, filtering, and data visualization capabilities. The reports can be refreshed in real-time or according to a schedule. They can also be published to dashboards where they can be made available to a very large audience.

Data visibility

Data democratization is a key benefit of the SSA platform, which provides users across the spectrum with increased visibility into their data. According to the Aberdeen Group, 83% of self-service users saw an increased visibility into business data[5]. It appears logical that when users are provided access to their data, they will dig more which will provide them with more data visibility and insight. When this happens, they can initiate improvements in diverse areas such as data collection, data cleaning, analysis, operations, performance, etc.

Insight without data scientists

The reporting and analysis requirements of most business users are fairly basic and straightforward. These can be satisfied through interactive reports that support simple functions like filtering, drilling, sorting, etc. The results provided through these operations are simple and can be understood by most business users without any significant explanation. A user who understands the business can easily figure out items such as revenue, costs, sales, inventory, returns, etc.

---

5 Self-Service Data and the Customer Connection, Aberdeen Group, March 2017

When complexity arises, business users may need the help of experienced analysts or data scientists. For example, they may be faced with questions like:

- What is the optimal mix of products to achieve the highest ASP (average selling price)?
- Based on available data from five different data sources, what is the customer sentiment index expected to be 6 months from now?

These are the exceptions rather than the rule, as most analysis by the majority of business users can be performed without the help of data scientists and experienced data analysts.

Access to diverse tools

A key reason that the self-service analytics concept has become popular has been the development of versatile front-end tools which are intuitive, interactive, collaborative, and very easy to use compared to the legacy business intelligence tools. These tools are very powerful and, in many cases, can fulfil the needs of the full spectrum of enterprise users ranging from casual users to experienced report developers and data scientists. In general, these tools perform three different core types of analytics:

- Descriptive: What has happened
- Prescriptive: What should happen now
- Predictive: What will happen in the future

Collaboration

In a self-service analytics platform, business users and the data scientists can collaborate to divide and conquer. The business users can focus on the core reporting and analytics needs, while leaving the heavy lifting to the data scientists and IT developers. Such a partnership provides optimal results. The initial analysis can be performed by the business users and, subsequently, technically advanced users like data scientists and developers can take their input and perform more sophisticated reporting or analysis.

Chapter 7

SSA BENEFITS FOR IT

Addresses legacy process shortcomings

In the legacy business intelligence systems, IT used to build the data warehouses as well as the datasets for the casual users and power-users. IT was required to develop them as it had the technical expertise to do so, which required knowledge of programming, ETL, SQL, etc. The business users did not have the technical skills and, therefore, just used whatever was served to them by IT. This process did not serve the needs of either the business or IT, which the SSA addresses.

Inefficient process is replaced

Employees in organizations, especially large ones, have been used to working in silos and with limited data. Whenever they needed access to more data sources or wanted the data to be structured in a certain way, they would reach out to IT, as it controlled and managed the platform's data. Depending on the IT request backlog and the complexity of the request, which could involve ERP systems or data warehouses, the time between request and fulfillment could be weeks or months. In many cases, by the time the request was fulfilled, the need would have dissipated due to the delay. A self-service analytics implementation gets rid of this inefficient process, which benefits both IT and the business.

Support requirement is reduced

With the introduction of tools that are easy-to-use, inexpensive, and powerful, business users have been empowered to develop their own datasets, reports, and dashboards. However, despite becoming self-sufficient to a certain degree, business users still require IT support. With SSA, the IT department's business support requirement shrinks considerably, since it no longer has responsibility for developing reports and dashboards.

This is one of the greatest benefits for IT because a major portion of their workload, which was consumed by report writing and dashboard development, gets offloaded. In the new scenario, business users have the ability to take care of their most common reporting and analysis needs, which enables IT to focus on more technical tasks.

Support levels can be tiered

The level of help will vary quite a bit from department to department because some are more technically savvy while others are not and, therefore, will need more support. For example, marketing and finance are more at home with technology than legal or human resources. Some departments have their own analysts, while others are completely dependent on IT. Therefore, the level of support will be in direct proportion to the department's technical know-how, analytics self-sufficiency, and level of independence.

Working with a single support model will be challenging and not be responsive to the needs of the business. Therefore, multiple support tiers may need to be created. For example, with one department, IT may create only the datasets. For another group, it may create the datasets and also develop complex reports and dashboards. However, for a very small group, it may need to develop datasets, reports, and dashboards.

IT becomes more responsive

An SSA environment allows IT to focus its effort on governance, security, data quality, and performance. Since the business no longer needs

IT for the tasks that they can perform themselves, their dependence on IT is reduced. The new sharing of responsibilities and lower support load enables IT to be more responsive to the business. Therefore, this scenario suits both IT as well as the business.

Permits focus on technical aspects

With the business performing work that was previously handled by IT, it frees up valuable resources who can perform more technical tasks. Instead of having to respond to requests from the business for developing reports and dashboards, IT can focus on the infrastructure aspects such as managing data sources, tools, ETL, data quality, security, governance, complex report and dashboard development, and performance.

Shadow IT security issue is reduced

In an SSA environment, governance and security become even more important as data, which is exposed to inexperienced business users, can be mishandled or get into the wrong hands. In such an environment, business users perform testing and provide the seal of approval. However, IT manages technical tasks such as data modeling, connecting to data sources, migrating reports and dashboards from one environment to the other (Development, QA and Production), etc. It is also responsible for managing permissions and authorizations.

With governance rules and protocols in place, and the looping in of shadow IT groups, data becomes more secure. This is a benefit for IT as they are ultimately responsible for ensuring that data is safe and no unauthorized access takes place.

Increased return on investment

The investment cost for data infrastructure and applications can be spread over a far larger business user population in a self-service analytics platform. Therefore, it provides a much better return on investment (ROI) for the enterprise assets.

Chapter 8

READINESS ASSESSMENT

The implementation of an SSA platform introduces significant changes for an organization. Therefore, an assessment should be performed to determine if the organization is ready for the change before an SSA program is implemented. The following sections describe the key indicators that should be evaluated to determine if an organization needs an SSA platform and, also, if it is ready for the changes that the program will entail.

Demand from executives

An organization is ready for an SSA program to be implemented if its executives have decided that they want a data-driven organization or have realized that shortcomings in analytics are negatively impacting decision-making. These top-down indicators are the best indicators of organizational readiness. When the top executives are the drivers, there will be fewer barriers and the probability of success will increase significantly.

Shadow IT groups

When departments start buying their own BI tools and developing their own data sources which are outside the IT-managed environment, they are effectively creating shadow (rogue) IT groups. When such self-service groups exist and multiply, it is a very good indicator that the organization is ready for a formal enterprise-level self-service analytics implementation. In such a scenario, the organization already has an experienced group of users, who can be very valuable when the enterprise program is implemented.

Reporting issues

The indicators of a serious reporting issue include a very big report backlog, departments upset by the report prioritization process which does not favor them, and new requests coming in at a faster rate than reports are being developed. Another indicator is that the existing data sources don't work for new report requests, which leads to the need for additional data sources or modification of existing ones.

Another reporting issue is that a report leads to more questions, which cannot be answered without developing another report that will need to go into the IT development queue. Also, when the report development cycle is long, the underlying data can change during that period. This can be frustrating for the users as it prevents timely or accurate analysis. If reporting becomes a big issue, it is a good indicator that points to the need for an SSA platform which can help eliminate the problem.

High demand for merged data

The analysis performed by business users can span multiple data sources, which can include internal as well as external sources. This data can be quite varied and include structured data, unstructured data, Big Data, Excel spreadsheets, data warehouses, data marts, databases, websites, etc. The need to merge or blend two or more of these data sources is fairly common. IT is typically too busy to respond in a timely manner and the business users may not have the skills or bandwidth to mash-up the data. This type of situation is a good indicator that the organization is ready for migrating to an SSA platform.

Demand for visual analysis

Decision makers need business intelligence tools, especially visual discovery tools, which can help them make decisions quickly and easily. Data visualization makes data analysis easier and faster, exposes problems at an early stage, provides early warning alerts, and indicates trends in an easy-to-understand format. Its use has been driven by the large volumes of data being collected and the widespread availability of BI tools at an affordable cost. When users start requesting more visualizations, it is a good indicator that users have started their migration from standard reporting to visualizations, which are better for data analysis and decision-making.

Prerequisites for SSA

Self-service analytics programs are very ambitious, which should not be implemented unless the organization is ready. Readiness evaluation should include a review of the critical components required to support the self-service strategy. The following are prerequisites for an SSA program to be implemented:

- Management wants to create a data-driven culture across the enterprise
- Organization wants to empower its users and enable data-drive decision-making
- Need exists for providing one version of the truth
- Regulatory and compliance issues require that shadow IT groups be incorporated in a governed environment
- IT is overloaded and resource-constrained
- Business users are frustrated due to unresponsive IT
- Current technology structure for reporting and analytics is not scalable
- Lack of tools that can meet the data and reporting requirements of diverse types of enterprise users

Readiness for onboarding

The following requirements indicate that a department or group is ready to be onboarded onto an existing SSA platform:

- Department want to develop reports without depending on IT resources
- Wants to obtain answers to specific questions that existing reports cannot answer

- Department data is insufficient for required data analysis and decision-making
- Department has skilled business users who understand the business as well as data
- Technical resource is available to manage the technical back-end requirements, such as connecting to data sources or developing extracts
- Department wants to share reports and analyses with other departments and vice versa
- Requires the ability to use standardized data definitions and metrics that are used across the enterprise

Chapter 9
PLANNING AN SSA PROGRAM

An SSA project is a very challenging project, which needs to be well planned and executed. It is more than a technology project, which also involves changing the existing culture to a data-driven culture. Before embarking on such a project, the process and organizational changes must be identified and planned for. The following sections describe the key tasks involved in the planning of an SSA program.

Understand the objectives

The introduction of any new technology or platform must be designed so that it meets the organization's business objectives and prepares its users for meeting them. Implementation must be preceded by careful planning, identification of the risks, and the development of a rollout strategy in order to ensure success.

The implementation planning for an SSA platform should be comprehensive and provide the tool(s) that meet the needs of the users so that they can develop their own reports and perform their own analysis. It should support customization so that it meets the specific needs of the business users and, also, enables collaboration.

Align initiative with organizational goals

The SSA initiative should be aligned with the organizational goals. The approach should be validated with measurable successful use cases. The value of an SSA approach should be confirmed by sharing its impact, providing specific linkages to successes, and providing specific benefits to the organization. This provides confidence in the selected approach, gets business support, encourages more users to get on board, and justifies continued support.

Obtain sponsorship

The first step in planning for an SSA project is to obtain business sponsorship. Getting the right project sponsor on board is important for many reasons. The sponsor is the key project champion who links the project, the business, and decision-makers. Additionally, the project sponsor ensures accountability for realizing project benefits, oversees the project, acts as the escalation point for issues and decisions, provides help when needed, removes roadblocks, and motivates employees.

Design the architecture

In this step, the SSA platform architecture is designed. SSA platforms are rarely independent of the existing BI and operations infrastructures. They will usually leverage those infrastructures, which feed them data from the enterprise data warehouses, OLTP databases, data marts, data lakes, Hadoop, and other data sources. After the SSA architecture is developed, the components required to be integrated, upgraded, or procured are identified.

Select the right tool

The foundation of an SSA infrastructure is the front-end tool, through which the users interact with the platform. If the appropriate tool is not selected, the platform will fail to attract and retain users. The tool should be selected after an in-depth evaluation of its current footprint (if any), compatibility with other tools, types of users expected to use it, types of analysis expected, security, governance, performance, features, and limitations. Key feature evaluation considerations include flexibility, ease of use, integration, visualization capabilities, and collaboration. The evaluation should also consider features that favor IT or the business, as well as shortcomings. The evaluation must be performed as a joint effort by IT and the business or there will be issues in buy-in.

Install the platform infrastructure

This task involves the installation of the various technical components of the SSA platform. The specific components to be installed depend on the existing BI infrastructure and what can be leveraged. Since each environment is different, the specific items to be installed can vary significantly. In general, the items to be installed can include data sources, data lake, data warehouse, data marts, OLAP, servers, ETL software, as well as multiple software packages including the front-end tool, monitoring tools, governance tool, master data management (MDM), training environment, etc.

Setup multiple environments

The SSA platform should be setup with multiple environments: development, quality assurance (QA), and production. This provides an architecture in which users can work without worrying about making mistakes, while ensuring that the production environment remains secure and free of content that has not been approved. These three environments provide a migration path, which ensures that only tested and verified objects are migrated into production.

Development is an exploratory environment in which users can experiment and prototype. They can develop or modify reports and dashboards, which can be discarded or moved to the next stage (QA). Sometimes, this environment is called the sandbox. In the development environment, experiments can be conducted with new data sources, modified data sources, models, etc., without compromising the platform integrity. This experimentation can lead to new insights. The governance is light in this environment and the users can perform their exploratory analysis with existing or new tools.

In the Quality Assurance (QA) environment, objects like reports and dashboards are tested to determine if they are satisfactory before being migrated to production. The user acceptance testing (UAT) is carried out by the business in the QA environment. After passing the UAT, objects are migrated to the production environment. Usually, the migration of objects from development to QA and from QA to production is handled by IT. End-users access the final approved SSA content in the production environment.

Setup a data governance program

The data in an SSA system must contain verifiable information, which will ensure that it can be used with confidence. A data governance program should be implemented, which will ensure accuracy, integrity, and data safety in the SSA platform. Any new data brought into the platform, especially ungoverned data, must be validated before users are permitted access. Governance benefits users as they are provided access to data definitions that are accepted across the enterprise, which reduces the issue of inconsistent reports and improves trust.

The data governance policy must be transparent and comprehensive, which ensures that only authorized users can access the system and data remains secure. However, it should not be a rigid barrier that prevents flexibility because that will turn away users. The success of the platform is, to a large degree, dependent on the flexibility of the governance model, which should be strict enough to prevent any issues pertaining to data integrity and security. It should provide users with flexibility that will enable them to easily perform analysis and data discovery, while maintaining rules that ensure safety, confidentiality, and data integrity for the individual departments and the enterprise.

Maintaining a governed system requires that it should be integrated into the operations. Implementing governance is not a one-time effort.

Vigilance should never be relaxed, governance rules should always be followed, and periodic audits should be conducted.

Define partnership responsibilities

An SSA program should be based on a partnership between the business and IT. Their collaboration can provide many benefits, such as identifying the correct data sources for import, defining user roles and their permissions, training, testing, data migration, etc. These two partners need to have close cooperation and trust.

The business and IT roles, responsibilities, and mutual expectations should be made clear at the start so that there are no ambiguities and issues during or after implementation. This will help avoid issues at various stages from implementation through support. The IT/business partnership should be developed within a framework that should be defined at the start of the program.

Develop onboarding processes

Just implementing an SSA platform's infrastructure and tools is not sufficient for success. The users interested in utilizing the platform will come from different backgrounds and have skillsets that will vary significantly from group to group and user to user. The data to be onboarded onto the platform will come from numerous sources and different formats. The complexity can vary depending on the data sources, type of analysis needed, etc. Therefore, there will be many challenges in onboarding data, departments, groups, and users onto the platform.

Processes need to be developed for onboarding departments, users, and data. The users should be provided guidance and help in joining the platform, getting started with the tools, training, as well as how to get help. The onboarding program should be scalable and automated as much as possible, as potentially thousands of users and numerous datasets may need to be onboarded.

Setup a Center of Excellence

Setting up an analytics Center of Excellence (COE) provides a smooth transition to self-service delivery. The COE should include resources from IT as well as the business. There are many benefits associated with a COE. These include creating a framework for adoption, helping understand the business needs as well as the data required to support them, providing skills and expertise, project management, IT governance, knowledge management, and best practices, as well as user support and management.

Chapter 10

STAKEHOLDERS AND TEAMS

Stakeholders

There are three key stakeholder groups in an SSA platform:

- Management
- Analytics team
- Users

These groups are described in the next three sections.

Management

In an SSA program, the most critical stakeholder is management, which can be at the enterprise or the department level. Though they form the smallest group of users, management users are the most important because of their position in the corporate decision-making hierarchy. Their requirements and reports are usually delivered via executive dashboards. The implementation project sponsor will typically be drawn from this group.

Analytics team

This team designs and implements the SSA platform. It is responsible for the SSA technical infrastructure including the environments (development, quality assurance, and production). The responsibilities of the analytics team include identifying technical requirements and installing databases, servers, software applications, and tools. They are responsible for identifying the data sources, adding and maintaining data sources, data preparation, and data modeling, as well as developing complex reports, views, and dashboards.

Users

These are the employees who use the SSA platform for their routine day to day work, as well as for any analyses that they may need to perform. The level and intensity of their interaction with the SSA platform will vary, with some using it only occasionally while others will interact with it for hours every day. Some users will just point and click on a report or dashboard to get the results they need, while others will perform in-depth analysis. Some users might filter and group data, while others might merge data from different sources and then perform some analysis. The level of interaction, as expected, will be dependent on the role such as casual user, data explorer, power-user, developer, or data scientist.

SSA teams

To enable, implement, operate, and provide oversight of an SSA program, a number of teams are required including:

- Steering committee
- Project Management Office
- Technology
- Governance
- Processes
- Training
- Communication

These teams are described in the following sections.

Steering Committee

The Steering Committee is an advisory body, consisting of stakeholders and subject matter experts, which oversees the program. Its objective is to ensure that the project aligns with the business objectives and steer the project to successful completion. The Steering Committee provides advice on important project issues and guidance regarding policies, procedures, financial aspects, resourcing, etc. Other tasks associated with the Steering Committee include:

- Approve program scope and approach
- Determine project outcomes
- Prioritize project steps and goals
- Monitor/eliminate project risks
- Resolve conflicts
- Set project timelines
- Monitor project progress
- Approve scope or budget changes

Project Management Office

The Project Management Office (PMO) is a centralized project management structure, which provides an approach to identify, prioritize, and execute a portfolio of enterprise projects aligned with its strategic goals. The PMO works with the program sponsors, Steering Committee, and IT to set the program direction and scope.

The PMO's typical responsibilities and tasks include managing the project portfolio, coordinating the activities of all program teams, establishing project methodologies, defining and maintaining process standards through a framework, providing tools and procedures, as well as setting and tracking change control procedures.

The PMO establishes standard performance measures (based on the enterprise objectives), ensures standardization, reduces duplication, leverages resources, and provides project tracking and reporting on progress, outputs, and outcomes. It also manages cross-project inter-dependency, provides project support through guidance to the project managers, tracks and manages risks and issues, and also escalates when needed.

In an SSA program, program management is a shared responsibility across multiple organizations, which include governance, business units, and the SSA program team (which includes representatives from IT and the business).

Technology

The technology team is responsible for installing and maintaining the SSA infrastructure, such as servers and tools. They are responsible for ensuring its security, user setup and management, and other related tasks such as transporting objects from one environment to the other. The technology team is responsible for the overall process of making data available to the SSA platform. The team's other responsibilities include:

- Data management, which includes managing and maintaining data repositories
- Ensuring high availability and performance for the infrastructure
- Ensuring that appropriate data access controls are in place
- Maintaining the data catalog
- Developing and maintaining SSA data services and processes
- Developing complex reports and dashboards
- Adding new data sources and modifying existing ones
- Providing connections to data sources
- Resolving technical issues for business developers and power-users

In some cases, a dedicated analytics group is setup within IT. In other cases, this group is part of the business.

Governance

The data governance team is responsible for ensuring that data within the SSA platform has been subjected to the appropriate data governance oversight and review. This team is responsible for managing the process that governs data assets, measures and metrics, as well as the dashboards. It also develops the end-user facing data governance services and processes for the program, as well as governance standards and templates. This team also provides the required training pertaining to data governance processes and requirements.

Processes

The process team is responsible for developing, documenting, and maintaining the processes required to operate and maintain the SSA platform, which include onboarding, licensing, training, governance, testing, data migration, issues management, user management, etc.

Training

The training team is responsible for preparing the users so that they are able to develop the skills required to perform their work in the SSA platform. The training team responsibilities include:

- Developing the training and documentation strategy
- Identifying and defining the SSA roles
- Developing role-based training
- Developing training documentation
- Delivering training through various methods (online, instructor-led, webinars, etc.)
- Performing training assessment and certification
- Developing and maintaining the training environment for hands-on training
- Developing help and best practices documentation
- Conducting user group sessions
- Managing the Learning Management System (LMS)

Communication

The communication team is responsible for communicating information about the SSA program to all stakeholders. Its key responsibilities include developing the communication strategy, designing and maintaining the program website, as well as performing outreach to departments, groups, and users in order to inform them about the program and any changes being introduced.

Chapter 11

CASUAL USERS

Need for different roles

An SSA environment can literally have thousands of users. Therefore, it is necessary that the roles they play and the information they need should be understood. Only after the roles are defined can the access tool be configured to reflect the role-based requirements. This ensures that a user only has access to the necessary functionality and data, while other features and data are restricted.

Benefit of grouping users

A benefit of user grouping is that it enables appropriate training to be provided, based on the users' different needs and requirements. Such segmentation, which avoids the one-size-fits-all approach, is a better approach for ensuring a successful implementation and providing the right skills to the users. Another benefit is that once the user groups are defined, they can be mapped to the information/data architecture.

User groups

The primary groups in an SSA environment are:

- Casual users
- Power-users
- Developers

How users utilize self-service

There are five distinct ways in which SSA platform users utilize self-service. These are:

1. Use
2. Analyze
3. Prepare
4. Store
5. Inject

Developing roles and mapping users ensures that every user is mapped appropriately. Depending on the role, a user may be allowed one or more of these five capabilities. For example, a casual user will only be provided permission for the first capability (use), while a power-user may be provided authorization for the first three items (#1-3).

Who are the casual users?

Casual users are the consumers of reports and dashboards that have been created by power-users and developers. These users need access to information in order to do their routine day-to-day jobs. The self-service analytics needs of these users can usually be satisfied by a simple or interactive report or dashboard. Typically, casual users can only access specific reports or dashboards relevant to their job function or department. However, they cannot develop or publish new reports or dashboards.

Work performed by casual users

A casual user's interaction with the front-end tool is limited to navigating and viewing a report or dashboard. Such a user may also perform subsequent tasks such as change field selections for time periods, perform drilldowns, re-format a report display, share a report with other authorized users, or print a report. Casual users can also provide requirements and feedback to power-users for creating or enhancing reports and dashboards. They may also participate in User Acceptance Test (UAT) activities for reports and dashboards prior to their migration to production.

Percentage and types of casual users

The percentage of casual users in an SSA platform varies by organization, though It is estimated that they make up roughly 90 percent of all users. Typically, they include executives, managers, and line workers. Such users have limited technical skills and their requirements are relatively modest. In most cases, dynamic reports and dashboards are sufficient to take care of their reporting needs.

Casual users are subdivided into two sub-groups:

- Data consumers
- Data explorers

Data consumers

Data consumers are the ultimate consumers of reports and dashboards. Their interaction is limited to operations like launching a report, filtering, drilling, sorting, and sharing. If permitted, they can download the results. This group accounts for about 60% of the users in an organization, with data explorers accounting for about 30%.

Data explorers

Data explorers, who are also consumers, work on existing content which may have been developed by a power-user or developer. However, their interaction with the platform has more depth compared to data consumers. Data explorers like to perform limited ad-hoc analysis, drilldown for more exploration, slice and dice, add or remove fields from an existing report, develop simple calculations or metrics, etc. In some cases, they can also create simple reports from scratch. Data explorers consume more than they create.

Data explorers provide support to their own department's data consumers. They also participate in user acceptance testing (UAT) for their department's reports and dashboards. They identify any changes needed for existing reports and dashboards and, also, identify new reporting and analysis requirements.

Chapter 12

POWER-USERS

Who are the power-users?

Power-users are skilled business users who require more flexibility and functionality for extracting answers for their business questions. When a business intelligence application is deployed, these users quickly become tool experts and start using it for ad hoc analysis. Within a short time, their expertise leads to their supporting the casual users. The casual users find them easier, faster, and more responsive to work with, for developing custom reports and views, compared to backlog-hampered IT. Power-users are critical for the success of an SSA program. They play a very valuable role and, in some ways, are an informal extension of the technical team.

Power-users have more technical skills than casual users (data consumers and data explorers). In addition to their business knowledge, they are knowledgeable about data and statistics. They are also familiar with techniques and tools like data blending, data modeling, querying, SQL, Hadoop, data lakes, and machine learning.

Work performed by power-users

Power-users perform data exploration, ad hoc analysis, and also modify or develop reports and dashboards. They are able to connect to new data sources, merge data sources, and prepare data. In some cases, they also code and develop custom scripts. Power-users also play an important role in UAT, where they take a leading role in testing and approving reports and dashboards before they are moved into production.

Other tasks that power-users perform include supporting their department's casual users, interpreting data, identifying new requirements or changes to existing reports and dashboards and, also, helping resolve issues for casual users.

Percentage and types of power-users

Power-users are subdivided into two sub-groups:

- Data analysts
- Data scientists

The precise functionality that a power-user may be authorized for varies from implementation to implementation. It is estimated that approximately 10% of users are power-users, with a breakdown of 8% for data analysts and 2% for data scientists. These percentages are approximations as these numbers can vary significantly by industry, organization size, and definition of power-user. For example, high-tech and financial services companies have a higher percentage of power-users compared to manufacturing or the food industry. Some estimates of the power-user percentage are higher for larger organizations.

Data analysts

Typical reports made available to casual users do not meet the needs of data analysts and, therefore, they frequently develop their own reports and dashboards. Their toolset includes SQL, statistics, scripting, and advanced Excel. Data analysts intimately understand their department's data and business and, therefore, are in a unique position to mine the data and provide insights that can benefit the business.

Data analysts are experts in working with and manipulating data. They can identify the data required for analysis, connect to it, and then prepare it for consumption by the front-end tool. In many cases, they use multiple front-end tools for their analysis, which can be very beneficial as it enables them to use the appropriate tool for a specific problem. The precise functionality that a data analyst may be authorized for varies from organization to organization.

Data scientists

These are power-users who have a very technical background that typically includes a degree in computer science or mathematics. Data scientists are at home with languages like R, Python, Hive, Java, SAS, and MATLAB. Their skills enable them to look at vey granular data from different perspectives, with different tools and languages. They are qualified to cover the gap that power-users cannot fill, due to their limited technical skills and lack of depth compared to data scientists.

Before SSA came into vogue, most of the complex analysis was performed by data scientists. However, due to the availability of advanced tools and technologies, more and more analysis is now being performed by power-users and citizen data scientists, who are self-service users without a formal background in data science. According to Gartner, a citizen data scientist is "a person who creates or generates models that leverage predictive or prescriptive analytics but whose primary job function is outside of the field of statistics and analytics."

Gartner has predicted that by 2019, the analytics output of data scientists will be surpassed by the output of business users with self-service capabilities. It has also been predicted that the number of citizen data scientists will grow to be five times the number of data scientists. If these two types of data scientists work in a collaborative manner, productivity will get a boost.

A sub-group that is included within data scientists is statisticians. Some experts consider them to be the third sub-category of power-users, in addition to data analysts and data scientists. Data scientists are characterized by their statistical skills, ability to work with machine-learning models, and ability to develop predictive models. Instead of a computer science degree, their background is in mathematics, operations research, economics, and other related fields.

Chapter 13

DEVELOPERS

Types of developers

Developers have a broad range of technical skills which are required to support the SSA platform infrastructure. The developer role is split into three sub-groups:

- Administrator
- Report developer
- ETL developer

In many cases, a resource may have the skills to perform tasks that span more than one role. For example, a user can act as a report developer as well as an ETL developer. Similarly, an administrator may also function as a report developer.

Administrators

A user in this role has the responsibility for the smooth operation of the SSA platform. An administrator installs servers, software, applications, and tools. Other tasks include monitoring the performance of platform components, managing users including the setup of users and roles, implementing and monitoring security, applying patches, performing upgrades, transporting objects between environments, managing data sources, etc. This resource is usually a part of the IT team.

Report developers

Report developers develop and modify reports and dashboards, monitor their performance, and provide connections to new and existing data sources. They also prepare data for use by the front-end tool, develop extracts, publish data sources and workbooks, add or create new dimensions and metrics, develop custom scripts, maintain the data catalog and metadata, and perform complex technical tasks. These developers also perform unit testing of reports and dashboards developed by power-users.

Developers in an SSA environment typically report to the business, though in some cases they work within the IT department. When working for the business, a developer is the department's point of contact with IT. Developers coordinate with IT to provide user access to the platform, migrate data and objects between environments, and confirm sign-offs prior to migration of objects to the production environment. They are also responsible for providing support to their assigned power-users for complex issues.

Developers with advanced skills can build custom analytic applications using programming languages, APIs (Application Programming Interfaces), and SDKs (Software Development Kits). Using their technical skills, they can embed and integrate with external applications.

ETL developers

ETL developers extract data from various data sources, such as data warehouses and databases, and then load them into the SSA environment where the data can be used by the front-end tools. As part of this effort, they identify the source data, perform data modeling, design data flows, execute and monitor ETL jobs, and also perform other related back-end tasks. These resources are located within IT.

Chapter 14

MAPPING USERS TO THE SSA ARCHITECTURE

Identifying the role

The training for users must be relevant for their role such as casual user, power-user or developer. If the training is not relevant, it will be ineffective and lead to frustrated users unable to use the platform and tools for their work. Assigning a role is also required for other reasons such as developing role-based documentation, user management, data governance, security, and collaboration, as well as access to reports, folders, dashboards, and specific tool functions.

A method used to help users in steering them to the right role is to develop a simple quiz, based on a few questions. Examples of questions for this purpose include:

- Will you be developing any reports or dashboards? A no answer will mean that this user is a casual user.
- Will you be connecting to new data sources? A yes answer will default the user to the developer role.

Gating questions can be used to exclude users from a specific role. Additionally, multiple-choice questions, based on tasks to be performed, can be used to determine the role that best fits the quiz taker.

SSA architecture areas

A self-service analytics architecture is based on a foundation that consists of four distinct stages:

- Staging
- Data Hub
- Business Views
- Applications

These stages, which contain different types of data, are shown on Figure 1.

Need for mapping

As shown on Figure 1, users with different roles access different components of the SSA architecture, where they perform different types of analyses. This leads to the requirement to map the business users, technical resources, and technology to different parts of the architecture, which is also referred to as the information supply chain.

When users are mapped to areas where they can be productive, it makes the platform relevant and also makes it easy for them to convert data and analysis into action. This capability is important, as just performing analysis without turning it into action is meaningless.

Benefits of mapping

The SSA architecture provides self-service capabilities for the information supply chain by enabling authorized users, without depending on IT, to import their own data into the SSA environment and, also, develop or modify their reports in response to their business needs or questions.

In traditional business intelligence, the architecture/information supply chain used to be off-limits for non-IT personnel. IT only permitted power-users to access a few data sources which, typically, included a data mart or a subset of the enterprise data warehouse. These data marts contained a limited amount of data, which was typically restricted to a few subject areas. Such data could not be used to provide an enterprise view or analysis. Very few users were allowed access to the crown jewel—the

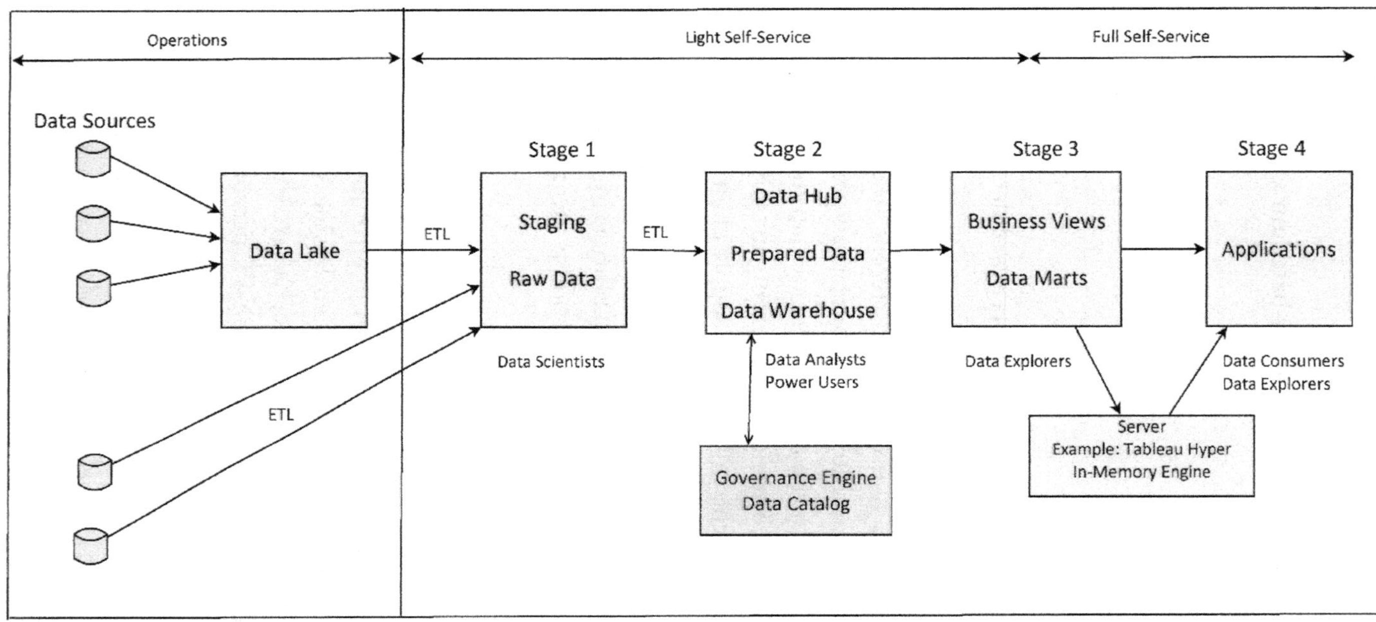

Figure 1: SSA Architecture

Enterprise Data Warehouse (EDW) or the data hub. An even smaller number of users were allowed to query data in the staging area.

In an SSA architecture, users can be mapped to and authorized for access to areas that were previously off-limits. When users are mapped to specific architecture areas, it provides many benefits. It makes the platform more secure, provides users the data and capabilities that they need, reduces the work that IT has to perform, and only users with the appropriate skills are allowed into environments that suit their backgrounds and capabilities. For example, casual users will not be permitted to access the staging area, work with raw data, or perform analysis that they are not qualified for. On the other hand, data scientists will be allowed to work in the staging area because they have the technical skills and knowledge required to query such an environment.

Mapping data analysts

Prepared data is available in the data hub area, typically in a data warehouse, where it has been transformed and organized into subject areas. This makes it ideal for the types of analysis that data analysts like to perform, especially slice/dice and drilldown operations using front-end visual discovery tools. Their productivity is improved when they are also provided with data preparation tools. Data analysts are mapped to the data hub, which is the area of interest to them.

Mapping data scientists

Data scientists tend to work with raw data which has not been modified, transformed, or aggregated. They can apply sophisticated models to such data for predictive analysis, machine learning, data mining, etc. Raw data is available in the staging area and is ideal for the type of work that data scientists perform. In this area, they can work in an unrestricted way. They can prepare data as per their requirements, transform data, merge or blend data, and use front-end and visual discovery tools for presenting the analysis through data visualization. Therefore, data scientists are mapped to the staging area.

Mapping data explorers

Data explorers are mapped to the business views, where they can develop new reports from scratch or modify existing ones. The data in this stage is typically served through data marts. In some environments, data explorers may also be allowed to use data preparation tools to merge existing data with other datasets, which may be local, external, or already available in the SSA platform.

Mapping data consumers

Data consumers, who are the end-users in the SSA platform, typically work with existing reports and dashboards that have been developed by others—data analysts, power-users, and data explorers. They cannot modify reports or dashboards, though they can perform some flexible analysis like drilldown, slice and dice, as well as change filters and selection criteria. After modifying a report, data consumers are usually permitted to save their custom views, which they can access later on. Data consumers are mapped to the applications stage, where they perform all of their work.

Chapter 15

UNDERSTANDING THE IT ROLE

Transitioning to a new role

As organizations begin the transition from a traditional approach driven by IT to a self-service approach enabled by IT and led by the business, IT leaders should redefine their role and develop a new framework. This means that past decisions supporting the core foundational components of a BI program—people, process, and platform—must be revisited.

In the new role, IT should become a strategic partner, which will change how the business perceives IT—as an enabler and strategic partner, rather than just a producer. It will provide the necessary infrastructure, processes, and guidance required to operate the SSA platform. The IT goal should be to maintain the appropriate balance of business autonomy and flexibility through self-service while ensuring the implementation of platform safeguards and stewardship.

Two IT options

There are two choices available to IT for controlling and managing the reporting and analytics needs of the enterprise. The first is to continue with traditional reporting along with its limitations, which will mean continued inflexibility, increasing backlog, increasing number of shadow IT groups, and failure to empower the business. The second choice is to accept change and empower the business through an SSA implementation, which will mean offloading some key responsibilities to the business users. This choice will deliver more benefits and value to the enterprise.

The transition from the traditional approach to the SSA approach, which is driven by the business and supported by IT, means that processes must be revisited and redesigned. This should encompass all three components: process, people, and platform.

Historical implementation approach

Historically, business intelligence projects have been driven by IT and managed in the same way as conventional IT projects. These projects have been implemented using the software development life cycle (SDLC). Waterfall has been the most popular implementation methodology, which is characterized by a long cycle that includes planning, defining requirements, analysis, design, development, testing and deployment. While this methodology provides many benefits, a serious issue is associated with this approach—lack of agility. The time from need identification to solution deployment is very long, which is not acceptable in the current business environment because it needs to be very short.

New implementation approach

In the modern business environment, implementation delays are unacceptable for a dynamic organization. Users in such organizations are very agile and, therefore, want to perform analysis quickly and iteratively. If a user executes a query, the results might lead to another query or the requirement for using another dataset. If the requirement is handed over to IT, their conventional development approach will be unacceptable.

This has led to the need for another methodology, Agile, which is characterized by short and iterative cycles. The results produced using the Agile methodology, though limited, can be rolled out to the users quickly and, if needed, refined or corrected iteratively. In this approach, business users are closely involved in every project cycle. This contrasts with the Waterfall methodology in which business users do not participate in the development phase—they just consume whatever is provided by IT. In recent years, IT has implemented most analytics projects using this methodology.

New IT role in development

In the current business environment, users demand flexibility, agility, and independence, which is available through self-service analytics. In the SSA platform, business users are actively involved in developing reports and dashboards, which was an IT responsibility in the past. In this new setup, business users have become independent of IT to a large degree and perform most of their own development. However, in many cases, IT continues to be responsible for developing complex reports and dashboards.

Though IT may or may not participate in developing reports and dashboards, depending on how the implementation is planned, its technical expertise and help with development is available to the business users.

Need for collaboration

In conventional business intelligence implementations, except during the requirements gathering phase, there is limited or no collaboration between IT and the business as the product is developed. However, in an SSA project, collaboration between the business and IT is critical for ensuring success. Both partners bring expertise to the table: IT understands the technology and data, while the users understand the business and how to derive insights. Their collaboration can ensure that the deployed platform meets the business needs and also drives adoption.

Partnering responsibilities

In an SSA environment, a partnership exists between the business and IT. The responsibilities of both partners should be clear and without ambiguity. Their roles and responsibilities in the various processes, such as report/dashboard development and data import into the environment, should be clearly specified to ensure successful collaboration. IT should engage with the users and understand the business, which will make it more valuable.

Is IT still relevant?

A misconception about self-service analytics is that by empowering business users, the need for IT is eliminated. However, just because the business users are provided self-service capabilities, does not mean that they can be completely on their own. A proactive IT department helps users who require guidance and assistance. It also remains responsible for the infrastructure, tools, governance, reliability, consistency, and performance, as well as platform and data safety.

Since IT has more technical resources, it should actively help in the development of complex reports and dashboards, as well as advanced analytics, such as predictive analytics. It should transition from its legacy role, where it formally serves the business by meeting their data and reporting needs, to a supporting role.

Chapter 16
DATA IN THE SSA ENVIRONMENT

Role of data

The lifeblood of an SSA platform is data. When this valuable resource is used effectively, it makes users more productive and also benefits the enterprise in many ways. Therefore, users must have an overall understanding of data and its characteristics. Depending on the role, some may need in-depth understanding, such as being aware of where the data came from, the transformations it may have undergone, period validity, etc. If SSA users want to be empowered, they must understand data, which is the foundation of SSA.

Data requirements

Any data that can be accessed by the SSA platform tools is considered "SSA data." The basic requirement to run an SSA platform is high-quality, accurate, reliable, and trusted data that can be accessed and analyzed so that insights can be derived easily and quickly. It should be available to the right users whenever they need it. This is a challenge because conventional BI tools have not kept pace with the fast growth of new data sources, which are of different types.

Enterprise data is characterized by huge volumes, which may also be changing very fast. Such data needs to comply with regulations and privacy laws, as well as the confidentiality needs of the business. Due to the complexity of the requirements and the availability of limited IT resources in many organizations, it is a challenge to provide data meeting these requirements to the business in a timely manner.

Data complexity

Data in an SSA environment can be quite complex due to its different characteristics and properties. The following are the key reasons that make data complex:

- Structured data: Can be sourced from machine generated data (POS data, sensory data, call detail records, web server logs), as well as human generated data (any data entered in a computer such as sales transaction data)
- Unstructured data: Can be sourced from video, audio, digital images, e-mails, PDF, Twitter, messaging, etc.
- Level of storage: Can include current detailed data, lightly summarized data, highly summarized data, aggregated data, and historical data
- Data source types: Can include data warehouses, data marts, databases, ERP systems, OLTP databases, Excel spreadsheets, clickstream data, etc.
- Converted or migrated data: Data pulled from other systems
- Big Data: Includes huge datasets that commonly used software tools cannot capture, curate, manage, and process within a tolerable elapsed time
- Cloud data: Can include any data that is stored off premises in the cloud

Reasons for data chaos

Data chaos is pervasive in organizations due to a variety of reasons. These include multiple systems and platforms that are not integrated, siloed departments, errors, poor quality, inconsistency, changing data sources, poor data modeling, data extraction issues, quality of technical resources, faulty processes, data management issues, etc. Frequently, users and analysts cannot find or easily access the data that they need. This has a financial impact and productivity loss for organizations.

Managing data chaos

To avoid or manage data chaos that results from data complexity, the first and foremost requirement is to have streamlined processes in place and become as agile as possible. This will help organizations to quickly respond to changes such as the need to merge datasets, introduce new data sources or types, and modify existing data. The following, if implemented and adhered to, will help in reducing data problems:

- Develop processes for ensuring data quality
- Enforce data governance
- Identify data owners and hold them responsible for their data
- Assign roles and responsibilities across the enterprise
- Provide data training

Data ownership

A common misconception is that IT owns the data because it is responsible for data safety and integrity and, also, provides access to it. The reality is that IT is only the data steward, while ownership resides with the business. Since enterprise data content varies significantly, the ownership is spread out in the business units and departments. For example, marketing data will typically be owned by the marketing department, while personnel data will be owned by human resources. Each dataset should have an owner, who will be responsible for deciding who gets access and, also, what each user can do once access is provided.

Benefits derived from the data asset

If results are expected from SSA users by just providing them access to data, the expectations are misplaced. These users have different backgrounds, experiences, and skills. Therefore, if they are expected to generate meaningful results, they must be well trained, especially in working with data.

With the exponential rise in the volume of data and, consequently, analyses that requires working with large and diverse datasets, it has become clear to organizations that data is a critical asset which can only provide value if it is utilized effectively. This value is not limited to any type of organization, process, or business function. The benefits extend to sales, marketing, finance, manufacturing, as well as other departments and groups. Data can be used to improve customer experience, reduce costs, increase revenue, identify patterns, improve reliability, improve visibility, and much more.

Chapter 17

WORKING WITH DATA

Locating the data

Providing a front-end tool is just the first step, which is not enough for a user to become productive. Users need to know which data is relevant and appropriate. Without guidance they will be lost considering the vast treasure trove of data that is available in a self-service analytics platform. Even experienced analysts can be confused due to the wide range of data sources that they can tap into, such as data warehouses, data marts, databases, extracts, spreadsheets, data lakes, cloud sources, and other external sources. After a data source required by a user has been identified, permission will be needed to access and work with it.

Connecting to the data

After a data source has been brought into the SSA environment and it is connected to the front-end tool, users can start developing reports and dashboards. However, when analysts want to connect to a new data source, they will need to specify the connection and enter the credentials required to access that data source. Data models and relationships developed with the newly sourced data should be checked and tested to ensure that they do not lead to poor performance.

Changing the data

The data available to most SSA users may be in a form that they can use as-is, without requiring it to be modified. However, in many cases, analysts need to confirm if the data is complete, accurate, validated, and current. If the data is inconsistent or redundant, it will need to be cleaned and prepared for consumption before any analysis can be performed.

The change operations performed on data in order to prepare it for use can include removing inconsistencies and duplications, transforming, merging, filtering, etc. Depending on the data quality and the specific changes that are required, the effort can be minimal or significant. Also, depending on the data security, integrity, and governance considerations, the changes may not be permitted in the original location. Such data may need to be copied to a secondary environment, where users can make the necessary modifications.

Consolidating the data

The consolidation task may be performed by the data analysts when they want to merge or blend data sources. In this step, depending on the types of analysis required, an analyst or developer may use table joins, create a new table, develop summary tables, or create aggregates.

Importing the data

Data needs to be imported into the SSA platform for a variety of reasons. This import can be on an ad-hoc basis, periodic frequency, or when the data changes or grows. Any data that is imported should conform to governance rules, be clean, and validated. The level of scrutiny will depend on where the data is being imported from, which could be an external source or a source outside the SSA platform but within the enterprise IT infrastructure.

Verifying the data

It is a best practice to verify data from different perspectives, such as completeness and accuracy, before starting any analysis. If the data is not verified, the analysis can produce unexpected and inaccurate results. The results produced can be verified through secondary means. For example, the totals produced by a front-end tool report can be verified by comparing it with the totals returned by executing an SQL query. In some cases, calculations can be performed manually or checked against previous reports, which were executed using a different procedure or tool.

Preparing the data

Data preparation is one of the most important steps in data analysis. It refers to the process of getting the data ready for use in a downstream process. The objective is to improve data quality, usability, and accessibility so that it can be consumed for analysis that can provide insight. An example is the combining of data from multiple sources to create a single view of a customer or product.

Data preparation process

Data preparation can be performed on a single data source. However, in many cases, a dataset's value and utility is increased significantly when it is combined with other related but disparate datasets. In such a case, the process typically starts by pulling data from different sources and then subjecting it to some or all of the following steps:

- Profiling
- Cleaning
- Enriching
- Formatting
- Parsing
- Merging
- Normalizing
- Transforming

As an example, take a case where data is merged from two sources, an Excel spreadsheet and a data mart, and an issue is identified in the "State" field. In one source, it might be stored in an abbreviated way (CA), while in the other it might be not be abbreviated (California). An analysis of the total sales will return two different totals due to CA and California. To resolve this issue, all of the California items will need to be changed to CA or vice versa. In another field, for example, there might be a combination of numeric and text values. These will need to be formatted or split into different fields before the data can be used.

Importance of data preparation

Data preparation increases the data's value, builds user confidence, and makes it easier to perform analysis on the prepared data compared to unprepared data. Self-service data preparation tools empower users who can perform many data preparation tasks without relying on scarce IT technical resources.

Who can prepare data?

In the past, data preparation was the exclusive responsibility of the technical users who were part of IT. However, the introduction of easy-to-use, point and click, self-service tools has provided data preparation capabilities to less technical users such as analysts and power-users. With the right business context and understanding of data, self-service data preparation tools can enable power-users to prepare the data themselves, which helps them avoid having to rely on IT.

Chapter 18

METADATA

What is metadata?

Metadata has been described as "data about data." Metadata is a key element in the data architecture which is critical for constructing, managing, and using data. It provides useful information for locating stored data. All data about the data, as mapped between the source and target systems, is resident in the metadata. The metadata includes a description of the data fields and tables, data types, and acceptable value ranges.

Metadata users

Metadata can be used by the technical resources such as IT staff, back-end developers, report developers, business analysts, managers, support staff, and administrators, as well as the governance team. It is also useful for power-users who frequently develop reports and queries. Metadata can also be useful in some cases for light and casual users, who may need additional details about the reports that they use.

Metadata and governance

Metadata is a critical element for maintaining the value of an organization's data and is at the center of any data governance effort. It helps identify, define, and classify data and, also, provides the means for users to manage the data content and its context. Metadata aids in accessing data and also enables data governance policy pertaining to data definition, usage, security, and lineage. Using metadata, data governance is able to exert control over data throughout its lifecycle.

Metadata management refers to activities which ensure that metadata is created, stored, and controlled as specified in the governance rules and policies. For data governance to succeed, metadata management has to be successful. It ensures context, without which data can be misleading and lead to poor decisions.

Metadata benefits

Metadata provides useful mapping information, such as where the data came from—legacy systems, files, databases, fields, etc. It documents relationships between data structures within or between databases. It also provides information about the data structure, tables and columns, as well as their definitions and descriptions. It is also the source of information about operations that were applied on imported data, such as:

- Selection criteria
- Filters
- Business rules
- Transformation and integration operations
- Cleaning
- Algorithms used to summarize and derive data

Metadata, which typically resides in a database or a formal repository, can provide many additional benefits. For example, it can be used to review how the business definitions and calculations changed over time and, also, provide a history of extracts and changes in data over time. An organization that understands the importance of metadata management will implement a solution which will ensure that all types of metadata, business and technical, is collected.

Enterprises that view metadata management as a critical resource are able to deal with issues and a changing business environment easily and quickly.

Chapter 19

DATA PIPELINING

The following sections describe data pipelining tools and the roles they play in supporting a governed self-service environment.

Role of pipelining tools

Data pipelining tools help manage data flow from the source to the target. For analytics, these tools typically perform the following tasks:

- Ingest
- Transform
- Validate
- Clean
- Merge
- Secure
- Format

These tools can support data capture and updates in batch mode, near real-time, as well as real-time. They support incremental data changes, which enable only the changed data to be brought in and, hence, reduce the data volumes to be loaded. These tools also capture metadata at every step, which enables data movements and transformations to be captured, which is very important when the data lineage is to be tracked and monitored.

Pipelining tool categories

The data pipelining tools can be broadly categorized into two types:

- IT-driven, which are IT oriented
- Business-driven, which are analyst/power-user oriented

IT-oriented pipelining tools

Extract, Transform, and Load (ETL)

This process is described in Chapter 23.

Streaming Engines

Streaming is a technique for serving or delivering content over the Internet. Streaming engines are tools that help capture real-time streaming data. Using a dedicated server with streaming server software can enable on-demand data to be delivered to the SSA platform.

Data Warehouse Automation

Data warehouse automation is the process for accelerating and automating data warehouse development cycles from analysis through documentation[6.] Its primary focus is the automation of every step in the data warehouse lifecycle, which leads to a reduction in the effort required to build, load, operate, and maintain data warehouses and data marts.

The benefits of data warehouse automation include shorter development time, flexibility, quality improvement, consistency, faster access to integrated data, rapid response to business changes, faster decision-making, increased productivity, as well as easier data warehouse modification and maintenance.

Data Lake Management

Data lakes have been increasingly used in BI systems, which has led to the setting up of systems and processes that can manage and govern data lakes. Data lake management software, which is still in its infancy, can perform

6 https://en.wikipedia.org/wiki/Data_warehouse_automation

a range of monitoring and management tasks. It can capture relationships, improve quality, and identify usage patterns. Its beneficiaries include data analysts, data scientists, and data stewards, who can utilize it to discover, catalog, and prepare data for Big Data analytics.

Data Virtualization

Data virtualization is a data integration technology that enables data to be virtually integrated across multiple, disparate sources, internal and external. This permits data required for analysis to be brought together in one virtual place, in real-time, where it can be accessed by front-end applications and tools. With this technique, no copying or moving of data is involved, which means access to data is easier and faster.

Business-driven pipelining tools

Data Preparation

Data preparation is the process for collecting and consolidating raw data into datasets, which can be used for analysis. Data preparation tools are applications that enable the preparation of these datasets. The tasks that they perform include cleaning, formatting, structuring, transforming, enriching, and merging. The key steps involved in the process include data analysis, creating an intuitive workflow, validation, transformation and backflow (which is used to replace the original dirty data in the source system).

Data preparation tools are a new class of BI self-service products that enable power-users to prepare the data themselves before conducting their analysis. These tools are quite versatile which can search for and access other external datasets and, as needed, clean, convert, and merge them before providing them to the analysts.

Data Catalog

A data catalog provides a single unified source of reference for data users to access and find datasets quickly and with confidence. For self-service analytics users, it enables data discovery and exploration. The data catalog is a very useful collaboration tool that aids in maintaining data quality and governance by providing users a single validated reference source, which eliminates inconsistencies and conflicting definitions. They can search for specific data and understand its metadata and lineage. A data catalog differs from a data dictionary in that it provides the ability to search and retrieve data.

Chapter 20

REQUIREMENTS

Need to focus on business outcomes

When the requirements for an SSA platform are being determined, the focus is usually on abstracts or specifics that are more technology-focused. The focus should be on the business outcomes, rather than explore all possible options or create "X" number of dashboards and reports for "Y" departments. The fundamental requirement and objective should be to develop the SSA infrastructure and processes. Once that is done, the required data and tools can be made available to the platform users so that they can independently develop most, if not all, of their reports, dashboard, and analyses.

Understand diverse requirements

An SSA platform is used by a diverse group of users who are located at different levels in an organization's hierarchy. They have different needs and responsibilities and, therefore, will have quite different requirements. These can include ad hoc analysis, integration of data sources, or simple point and click report execution. In such an environment, a one-size-fits-all approach will not work. Before such a platform is designed and implemented, the needs, expectations, and skills of the users expected to use the platform must be determined. Unless the user requirements are implemented, the platform will be unable to meet the needs of its users.

Key user requirements

The following sections describe the most common user requirements, which will satisfy the needs of most SSA platform users.

Modification of Reports and Dashboards

Self-service analytics should provide business users the ability to perform filtering, visualize key metrics, and flexibly change reports or dashboards so that analysis can be performed from different perspectives. This is a basic requirement that an SSA tool must support.

Development of Ad hoc Reports and Dashboards

The tools provided to the SSA users should, through templates and reusable objects, enable power-users to create ad hoc reports and dashboards which can be shared with other users, without requiring IT to be involved unless new data sources need to be added.

Integration of Private and Local Data

The ability to integrate local data, such as Excel spreadsheets or flat files, into reports is an important requirement that must be supported by the front-end tool. This ability enables business users to extend their analysis and not be limited to a specific dataset.

Development or Modification of Data Models

The self-service tools should enable advanced users, like power-users and business analysts, to modify or develop data models independently. Setting up different environments, including a sandbox, is ideal as that can enable such users to experiment without risking the production environment.

Avoid vague requirements

Developing good requirements is essential for ensuring that any SSA project is well planned and executed. Requirements should be precise and not vague or all-encompassing. For example, at one analytics implementation, the executives specified that the dashboard should contain all metrics—an impossibility. The project spent 15 million dollars over two years and failed, which led to many senior project executives losing their jobs.

The requirements can be gathered using a number of methods such as Joint Application Development (JAD) sessions, workshops, written questionnaires, and interviews (in groups and individually). Asking the right questions is very important as it can help extract very valuable information. Where necessary, the justification for a requirement should be elicited and documented.

Information to be gathered

The objective of the questions should be to understand the business requirements which, subsequently, can be mapped to the type of data and analysis that the users will perform. If the requirements are vague or not understood, it will ultimately cause the requirements to be revisited, which will waste precious time.

The key information to be gathered during the requirements gathering process pertains to processes, users, data, analysis, frequency, distribution and collaboration, as well as training. These are described in the following sections.

Processes

The information to be gathered pertains to the processes that will be impacted, processes which can be leveraged or modified, and processes that will need to be developed.

Users

This includes identifying the expected users, number of users in each role, types of actions and analyses they will perform, and the data each user type will access. Additional information to be determined includes identifying users who will develop and maintain reports and dashboards, as well as the number of licenses required based on the user count and roles.

Data

This includes determining the known data sources, potential data sources, data ownership, as well as resources responsible for data integrity and data governance. Additional information to be gathered includes determining the requirements for extracts or live data, identifying users/roles permitted to download data or export to Excel, historical and aggregate requirements, as well as archiving requirements.

Analysis

This includes determining the type of analysis to be performed by each user role, history needed to support the analysis, and snapshot requirements. Additional information includes determining the user types who will be provided interactive capability, such as slicing and dicing as well as drilldown to the lower level granular data.

Frequency

This includes determining the frequency of the updates for reports and dashboards, if and how the reports will be delivered (when automated), as well as the frequency of data refresh.

Distribution and Collaboration

This includes determining how users will access reports and dashboards, access method (such as via the cloud, intranet or web), as well as the ability to distribute reports and analyses via e-mail.

Training

This includes determining the training requirements for various user types, delivery modes (online or instructor-led), as well as potential vendor and external training.

Chapter 21

SSA ARCHITECTURE REQUIREMENTS

Need for an architecture

Defining an architecture is important because there will be data chaos without it. When such a structure is provided, it enables authorized users to find and use data while, simultaneously, prevents access to data to those who are not authorized. Reducing the time to insight is a key driver for the SSA platform, which needs to be built on a sound architecture.

A misconception exists that self-service means freedom from constraints within the platform. The reality is that by its nature, any architecture is structured and constrained. Therefore, even a self-service platform must be based on a solid infrastructure foundation—an architecture that enables it to function, even while providing flexibility and governance to its users.

What the architecture provides

A well-architected self-service analytics platform enables data to be pulled from various data sources, after which it can be analyzed for data-driven decision-making. Such an architecture provides business users the flexibility to work with data and produce meaningful actionable output, while allowing IT to manage it. In such an architecture, the relationships between various components should be well defined and, also, the front-end tool must be integrated with the underlying data management architecture.

Factors impacting architecture

The data landscape has changed significantly in recent years due to the exponential rise in data volumes, more data sources, introduction of unstructured data, access demands of users who number in the thousands, demand for access to additional data at short notice, access to real-time or near real-time data, and development of front-end tools that have provided analytics capabilities to the masses including those who are not technology savvy. All these factors have impacted the architecture and infrastructure requirements.

Legacy BI process and platform shortcomings

Historically, IT had complete control over the data, despite ownership residing with the business. After capturing the data, IT stored it in data warehouses, data marts, and databases. When the business users made a request for a query, it was put in the IT development queue. Subsequently, IT developed a report and delivered it to the business users. If the delivered report required additional data that was currently unavailable, IT would identify the data source and then bring it into the data warehouse or data mart after carrying out any needed cleaning, transformation, and validation tasks.

This was a slow and inefficient process for the users, who would have to wait weeks or months for the requested report to be delivered. If the revised report had an issue, it would again go back to IT, leading to an additional waiting period. In many cases, when a report was delivered, it was already outdated, which resulted in delayed decision-making or missed opportunities for the business.

The processes and architecture supporting legacy business intelligence platforms are inadequate for SSA platforms. Therefore, the basic requirement of the SSA architecture and processes is to provide the foundation which will support the new SSA platform.

Data process requirement

To implement a self-service analytics model, it is imperative that the data process be well understood to avoid data chaos and enable meaningful insight. The data process and the quality of the data architecture will impact the quality of the reports, dashboards, and analyses that are produced. Therefore, designing a sound data process and architecture is a critical requirement for developing the SSA architecture.

Agility requirement

The limitations of the legacy process have led to the popularity of the SSA platform. This new platform is user-friendly with a flexible architecture that can respond quickly to changing business needs. For example, if a new data source needs to be added to the platform, in most cases it can be done quickly and easily by a power-user without getting IT involved. The platform architecture should enable users to quickly develop their own reports and analysis and, also, collaborate and share their results with other enterprise users. Therefore, providing agility is a key requirement for an SSA platform.

Business and IT requirements

The platform to be implemented should meet the requirements of the business as well as IT. Their partnership cannot thrive if the platform does not meet the needs of both partners. The basic IT requirement is that the data and infrastructure should be safe, while the business wants a platform which is agile, supports flexibility, and is based on data that can be trusted.

The tools to be selected should be user-friendly and have features that the business desires. In contrast to the way in which BI tools used to be selected, which was by IT with very little or no input from the business, the SSA front-end tool should be selected after evaluation by both IT and the business.

Chapter 22

PLATFORM ARCHITECTURE

Sections of an SSA architecture

There are three key sections in the SSA architecture, shown on Figure 1, which are based on how data is managed in the platform. These are:

- Operations
- Light self-service
- Full self-service

Operations

In this section, the analytics platform is fed by traditional business applications, operational or transactional, such as an Enterprise Resource Planning (ERP) system like SAP or a Customer Relationship Management (CRM) system like Salesforce. These have now been joined by non-traditional systems like Big Data, social media (like Twitter and Facebook), Internet of Things (IoT), and other machine generated data which can be in structured as well as unstructured format.

The data sources can be fed directly into the staging area or via a data lake, where data from various data sources is dumped so that it can be available for downstream use. This area is usually managed by the database/back-end administrators.

Light self-service

This is the playground of the data scientists, data analysts, and power-users. They need more flexibility in terms of their ability to execute ad hoc queries and work with raw as well as prepared data. Such users do not want to be limited by data that has been loaded into the platform by IT. They may want to introduce new datasets from local or external sources, as well as other enterprise data sources.

Data scientists work in the staging area, where raw data is available to them. Data analysts and power-users perform most of their work in the data hub, where prepared data is available for their use. The data in this area, which is often served through a data warehouse, is fully governed.

Since power-users frequently need to experiment in a more flexible environment with minimal constraints, they can import data from internal or external sources. In this area, they can merge data from multiple sources and perform analysis in relative freedom, which is not possible in the full self-service environment. When testing is completed successfully in the light self-service environment, their reports and dashboards can be ported to the full self-service environment where they become available to the casual users.

In the light self-service environment, data analysts and power-users prepare the data through various operations like cleaning, transforming, and merging data. They also explore and analyze the data, frequently visualizing the data. This is also the place where they can develop or modify new models, as well as test and validate calculations, business logic, and rules.

Full self-service

The full self-service area is the environment where most of the work is performed by the vast majority of users—the casual users. Their typical tasks include executing their reports and viewing dashboards in order to monitor and analyze business performance and relevant metrics. Their work can include performing limited ad hoc analysis, using functions like slice and dice, filtering, grouping, etc.

The full self-service environment is a fully governed area where the data is fully curated and ready to use. Casual users are not interested in importing or manipulating data for their limited and relatively simple analysis. Since the data available to them is clean, accurate, validated, and fully governed, it is trusted and can be used without any concern about its quality or accuracy. Wayne Eckerson defines this environment as Silver Service because of the way data and content is provided to its users.

SSA conceptual architecture

Figure 1 shows the conceptual architecture for a self-service analytics platform, which indicates:

- How data flows in the SSA environment
- Where the users interact with it

In practice, variations of this architecture are implemented. For example, the data sources might be far more complex and include data lakes, Hadoop, clickstream data, social media, etc.

Conceptually, as shown on Figure 1, an SSA architecture consists of the following stages:

- Stage 1: Staging
- Stage 2: Data Hub
- Stage 3: Business Views
- Stage 4: Applications

Figure 1 shows the self-service analytics process divided into four stages, which are described in the next four sections.

Stage 1: Staging

This is the area where data is brought into the SSA platform by pulling data from various data sources. Using ETL, data is extracted from the data sources that run the business and, if required, external sources.

Typically, many of the data sources already exist in the IT-managed environment and can include data lakes, relational databases, data warehouses, clickstream data, etc. Data in this stage is accessed and analyzed by data scientists and developers.

Stage 2: Data Hub
This is the area where, using ETL, the required data is pulled in from stage 1. The data is extracted in the original format or after undergoing some transformations and, typically, stored in a data warehouse. This data, which is clean and validated, is utilized by a small percentage of the platform users—the data analysts and power-users. In this stage, data governance kicks in and the platform starts to interface with the data catalog.

Stage 3: Business Views
This is the area where data is pulled in from stage 2 and typically made available through data marts, where the modeled data is organized by subject areas. This area supports analytics performed by the data explorers. Business views can be physical or virtual business models.

Stage 4: Applications
In this area, applications such as dashboards, reports, and visualizations are developed using the data available in stage 3. The primary users in this area are the data consumers and data explorers.

Architecture Variations
The following are some common variations of the SSA architecture shown on Figure 1:

- Stage 1 (Staging): This is shown in two separate steps: Landing and Staging. When data is brought into the SSA environment from the data sources, it is dumped into the landing zone. From there, after some validity checks and transformations, it is moved into the staging area.

- Intermediate server: In this scenario, an additional server is added between the business views (stage 3) and the applications (stage 4). An example is a Tableau server that sits between the two, as shown on Figure 1. With its hyper in-memory data engine, it can boost query speeds and extract datasets significantly faster.

Self-service workflows

A self-service analytics platform consists of workflows that are iterative and bi-directional. These workflows include developers and business users, who work in tandem. They provide benefits including:

- Requirements can be changed, added, or tweaked iteratively
- Incremental changes can be tested and implemented
- Prototypes can be developed, tested, and rolled out in an iterative manner
- Data to be consumed can be refined
- Curated data can be quickly refined and enriched by the business users

There are two types of workflows:

- Source to target
- Target to source

Source to Target

The normal flow in a traditional business intelligence environment is from the data source to the target. The flow is:

Curate (by IT) > Create (by IT) > Consume (by Casual users)

At each step in the process, data undergoes processing, which refines and curates it as it gets closer to the target where the business users are located. In such a platform, IT is responsible for and manages the data curation process. Most of the report and dashboard development is done by IT, which may be supplemented by the power-users. This process places a huge load on IT, which leads to a report backlog. Since users cannot use the data when they want it, it causes frustration and impacts their productivity.

In a self-service analytics platform, the flow changes to:

> *Curate (by IT and Power-users) > Create (by Power-users) > Consume (by Casual users)*

In the SSA environment, IT curates most of the data. However, power-users are also provided the authorization to curate data for their own use. They can access enterprise data, local data on their own desktops, as well as shadow IT systems data. They can work in an iterative and incremental manner, which enables them to prepare the data as needed. After their data is prepared, power-users can analyze it using data discovery or data visualization tools. They can then share their analysis with other department or enterprise users. Subsequently, as needed, they can also act on the analysis or perform additional analysis.

Target to Source

In the reverse workflow, the process is initiated by the business users. After the casual users, data consumers, and data explorers have used the reports and dashboards developed by the power-users or developers, they will be in a position to identify shortcomings as well as potential enhancements. These are then shared with the power-users who modify the existing reports and dashboards and, if needed, develop new ones.

If the casual user's request requires changes to the data, the power-user will forward the change request to the IT team, which will prepare the required data. This data will then be made available to the power-user, who will develop the required report or dashboard. After development, the power-user will share the modified report or dashboard with the department's casual users. This process can continue iteratively.

Chapter 23

ARCHITECTURE COMPONENTS: SOURCES & ETL

The following sections describe some of the data sources that are used in an SSA platform.

Data sources

Massive volumes of data are generated every day by individuals and organizations. The data generated includes structured, semi-structured, unstructured, as well as human and machine generated data, which can be captured and stored in real-time or in batch mode. This data is stored in various data structures, which become the data sources for an analytics platform.

The most popular data sources feeding an SSA platform include data warehouses, OLTP databases, data lakes, ERP applications, CRM applications, social media, e-mails, text messages, help desk tickets, as well as clickstream and streaming data. The following sections briefly describe the most commonly used data sources used in an analytics environment.

Databases

The most common type of structure used to store data generated by business operations is the OLTP database. Despite the growth of cloud platforms, on-premise OLTP databases are the backbone of most IT infrastructures. Therefore, it is to be expected that they are also the most common type of data source feeding SSA platforms.

Data Warehouses

For reporting and analysis, the most widely used data structure is the data warehouse. It is a large analytical database, which is populated from a variety of source systems that typically run the business. A data warehouse, which contains subject-oriented and integrated data, can contain very detailed data, granular data, as well as aggregate data.

Data Lakes

A data lake is a centralized data repository where data is stored in its original format. It can contain both structured and unstructured data. In a data lake, there is no hierarchy or organization among the individual items.

Data lakes contain data from all traditional data sources as well as social media, websites, IoTs, sensors, machines, etc. They support all data types and schemas. The stored data, which provides flexible access, can be used for querying, reporting, data visualization, analytics, machine learning, and Big Data analysis. The usage of data lakes, which are becoming popular as enterprise repositories, has been boosted by cloud-enabled platforms like Amazon Web Services (AWS), Azure, and Google Cloud Platform (GCP). Data from these platforms can be easily loaded into an organization's data warehouses and data lakes.

Data lakes help enterprises manage the massive volumes of data that they generate. They often serve as the foundation for self-service analytics. They are ideal for machine learning applications because so much varied and valuable data is available on a single platform. Despite their potential, data lakes have not been mined for analytics as widely as expected because it is not easy for users to determine what they contain, difficulty in identifying data owners and, also, lack of trust in data from an unknown and unverified source.

Unstructured Data

Unstructured data refers to information that is raw and unorganized. Such data is not structured in a pre-defined way, has no identifiable internal structure, and lacks a pre-defined data model. Its examples include e-mails, PDFs, spreadsheets, digital images, video and audio files, social media (Twitter and Facebook), messages, etc. The lack of structure makes compilation a time consuming and laborious task. When unstructured data is analyzed, it can provide insights that structured data cannot and, hence, can prove very valuable.

Semi-Structured Data

This is a form of structured data that does not reside in a relational database. However, it does have some organizational properties that make it easy to analyze, such as tags or other markers. It supports hierarchies of records and fields within the data, which enable analysis. Examples include XML data and JSON.

APIs

An application programming interface (API) is a software intermediary, which enables two applications to communicate with each other by specifying how the software components should interact. It is a set of tools, communication protocols and routines that are used for building or integrating software applications. For example, an API can be used to integrate the organization's data with third-party external application data, such as weather or demographic data, which may be stored in a cloud data warehouse.

ETL process

Data Acquisition Process

To bring data from the source systems into the SSA platform, it needs to be subjected to the extraction, transformation, and loading (ETL) process. The extraction step involves pulling data from the source system, the transformation step involves subjecting the data to a number of operations, and the loading step involves physically placing the extracted and transformed data in the target database.

The ETL process is an important and complex task, which is handled by IT. It is a challenging task because it involves complex operations for cleaning and integrating data, of significantly varying quality, which can be stored in many systems and formats.

Extraction Step

The first step in the ETL process is extraction, which is used to pull data from the source system. The source data can be in different formats such as VSAM, ISAM, relational, and flat files. The extracted data is parsed to check if it conforms to a pattern or a specific structure. The techniques for performing data extraction vary and are influenced by many factors such as type and structure of the source data, availability of tools, extraction frequency, exception handling, etc.

Transformation Step

In the second step, transformation, the source data is subjected to a number of operations that prepare it for import into the target database. For this task, integration and transformation programs apply business rules or functions to the extracted data, in order to reformat and modify the structure and data elements. In some cases, very little or no manipulation is required. However, in many cases, a range of operations may be required on some or all of the data. The operations performed may include summarization, de-normalization, filtering, joining, translating values, encoding, sorting, aggregating, calculating a new derived value, and pivoting.

Loading Step

In the third step, loading, the extracted and transformed data is physically moved into the target database. The loading process and the volume of data it has to handle varies according to business requirements. For example, some functional areas may require data to be added once a day, while others may require real-time or near real-time data loading. Other factors that impact loading include the period for which the data is required, data granularity, and aggregate requirements.

The volume of data required determines the time it will take to complete the ETL process. Depending on the data volume, the loading time can vary from a few minutes to many hours, which can impact the frequency and type of analysis that can be performed.

Specific ETL Tasks

The specific tasks that an ETL tool may have to perform range from:

- Converting data: Changing the incoming source data into a unified format and definition
- Deriving data: Applying a mathematical formula to a field in order to create a brand-new field
- Filtering data: Screening unwanted data from source data files
- Integrating data: Merging files from different data sources and platforms
- Summarizing data: Combining tables
- Selecting data: Selecting and loading data based on triggers

ETL Challenges

Getting data into a single target storage area is an important and complex task, which is time consuming. The challenges the ETL task faces includes dealing with multiple complex environments, data sources with different architectures and structures, poor data quality, poor data model design, complex transformation and cleansing rules, conflicting business definitions and rules, historical data conversion issues, and lack of procedures for resolving inconsistencies, as well as obsolete and incompatible tools and technologies.

ELT Process

A variation of the ETL process is ELT, which changes the sequence of steps to extraction, loading, and transformation. The ELT sequence is applied when a data lake is used in the process.

Chapter 24

ARCHITECTURE COMPONENTS: STORAGE

Intermediate data storage

Businesses use a variety of data source types which include relational, structured, unstructured, and semi-structured. After the data from these sources has been manipulated and transformed into a common format, it can be used by the BI and analytics tools to develop visualizations, reports, and dashboards.

The data pulled in from a data source is not imported directly into the target storage area. It needs to be validated and may also need to undergo a few operations before being loaded into the storage area. In many architecture diagrams, this is shown as a single component (Staging Area), while in others it is shown as two separate components (Landing Zone and Staging Area). These are described in the next two sections.

Landing Zone

A landing zone is an intermediate storage area, which sits between the data source and the staging area. When raw data is loaded into this area from a data source, it is stored in its original form. The landing zone can be a relational database, Hadoop, or other type of data storage structure. Typically, only IT has access to the data in this area.

After the data has been loaded into the landing zone, IT checks the load for completeness and quality. If any major issues are discovered, the loading process is repeated. IT also cleans the data and rectifies errors before moving it from the landing zone into the staging area. Data cleansing is based on implementing rules and standards for acceptable data quality levels. Since this layer contains raw data, data scientists prefer to work here before it is transformed or aggregated.

Staging Area

The data staging area is basically a temporary storage area where data is transformed and enriched before it is sent to the target storage (data hub), which can be a data warehouse, an operational data store (ODS), or other type of repository.

In the staging area, data is organized for downstream distribution and sharing. Operations carried out in this area include creating metrics, summarizing, and the merging of base objects into linked subject area tables. For example, to enable easier analysis, all data associated with a product might be merged into a single view and stored in a wide, flat product table. Similarly, customer information might be merged in another table.

Other tasks that may be performed in this area include de-identification and masking to protect sensitive fields, standardizing naming conventions, and creating base objects at the lowest granularity.

Storage

Data from the staging area is loaded into the data hub, where the business users can start working with the data. Loading is done through the ETL process. For storage, cloud data warehouses have become quite popular in recent years. These are highly scalable, cloud-based database platforms which enable, when needed, easy and quick extension of computing and storage capabilities. They frequently use massively parallel-processing (MPP), which provides the ability to perform complex analytical queries significantly faster. The benefit of a cloud data lake is that data is only transformed when the user is ready to process the data.

Data can also be stored in structures that are more suitable for reporting, which can typically involve duplicating and de-normalizing data into a classic star-schema. OLAP cubes are widely developed and used for multi-dimensional analysis.

Chapter 25

OTHER ARCHITECTURE COMPONENTS

Business views

Data in the data hub is transformed by the business views into data objects, which can be used for analysis by specific groups of users. These views are developed by business analysts, data analysts, or developers. They are implemented in a database or in a semantic layer, where they are presented in business-friendly terminology that users can relate to, rather than in cryptic technical names. A variety of tools are available for developing the business views, which include business intelligence tools, OLAP cubes, database views, data virtualization tools, custom tools, etc.

Data analysts use metrics and dimensions that are exposed to them to develop reports, dashboards, and analytics applications. They can develop new metrics, add new dimensions, develop calculated and derived fields, and modify existing reports and dashboards, or develop new ones. Users with advanced developer skills, such as JavaScipt and SQL, can leverage APIs offered through BI tools to develop custom applications and workflows.

Applications

This is the final layer in the SSA platform where data consumers and data explorers interact with reports and dashboards. The work performed in this area includes both traditional reporting and analytics. Reporting is the most widely utilized and user-friendly component of the SSA platform, which touches the maximum number of users. These users interact with the applications through a user-friendly graphical user interface (GUI). For most users, interaction is via point and click, through which they access reports, data visualizations, and dashboards that are made available to them.

In this area, users are exposed to the modern reporting and data visualization tools like Tableau, Power BI, Qlik, Looker, and other similar tools. They can perform a range of activities including traditional enterprise reporting, ad hoc analysis, and analytics, as well as collaboration.

Most users are satisfied with traditional reporting. However, a small but important segment of users use the available tools for analytics, which provides them with more powerful insight into what is happening. Using the available data, they analyze current trends and also infer or predict what will likely happen. For example, these users may predict the churn rate or forecast the customer sentiment index, which is a good predictor of whether the company will lose its customers.

Environments

The SSA platform should support three different environments:

- Development
- Quality Assurance (QA)
- Production

Power-users and developers start their work in the development environment, which is their playground. In this area, they develop new objects (reports, visualizations, and dashboards), test modified objects, test the impact of adding new data sources, and also test new connections. They also perform experimental analysis without fear of breaking something. Once they are satisfied, the platform administrator or authorized developer migrates their objects to the Quality Assurance environment.

In the QA environment, power-users and selected users test the changes that were implemented in the development environment and migrated to the QA environment. In this environment, metrics and analyses are certified and approved as per the regulatory and governance requirements. Formal user acceptance testing (UAT) is conducted in the QA environment. Objects that pass UAT are migrated to the production environment, where they become available to the end-users.

Chapter 26

GOVERNED DATA FLOW

Information supply chain

The foundation of an SSA platform for analytics processing is the information supply chain, which maps the data flow from the source system to those who utilize it—the end-users. Figure 1 shows how the users and technologies are mapped to the information supply chain:

- Staging (Stage 1): Data scientists
- Data Hub (Stage 2): Data analysts and power-users
- Business Views (Stage 3): Data explorers
- Applications (Stage 4): Data consumers

The supply chain encompasses two distinct areas:

- Light self-service
- Full self-service (also called Silver Service)

In these two information architecture areas, work is performed by different types of users. The tasks they perform depend on their roles and their reporting and analysis needs. In the light self-service area, data analysts have the flexibility to work with the available data, as well as with data that they can import to supplement what is already available. On the other hand, those who work in the full self-service area, the data consumers and data explorers, are restricted to work with the data that is already available in the platform and for which they have been authorized.

A well-designed, comprehensive, and flexible information supply chain will ensure the smooth flow of governed data, facilitate self-service tasks, and provide easy access to the data users need, which reduces or eliminates their need to obtain data outside the governed system.

Information supply chain characteristics

IT is responsible for the various applications and technologies used to implement the information supply chain. These technologies include relational databases, data warehouses, cloud data warehouses (CDWs), Hadoop, data lakes, NoSQL, Spark, etc.

The architecture should be flexible as, in many cases, it has to support bi-directional flows. For example, a data warehouse can feed the information supply chain or, conversely, be fed by the information supply chain. When such a flexible platform is implemented, it can easily support diverse requirements pertaining to data or analytics.

Workflows

In a dynamic organization, workers will rarely be satisfied with the data they are provided and, therefore, they will always be making demands for more data. Hence, IT will always be adding more data to the SSA repositories. The issue is that they are always inundated with more requests than they can handle in a timely manner. A well-designed SSA platform will help alleviate this issue by incorporating self-service workflows.

A well-designed SSA platform's architecture supports workflows, which are iterative and support collaboration among its diverse user groups including casual users, power-users, developers, and IT. These workflows help produce data that is governed and available to the authorized users. They also enable the development of reports, visualizations, dashboards, and analyses. Since the platform can include both governed and non-governed data, different workflows can be used for handling the different processes associated with them.

Impact of new tools on information supply chains

Hadoop

Hadoop is a powerful new tool that has reshaped the way information supply chains are architected by enterprises. It can be used in conjunction with the existing infrastructure and modern BI tools to provide new methods for data storage, processing and management. The development of MPP databases, along with Hadoop and NoSQL, has led to levels of processing and analysis that were unthinkable even a few years ago. In just a few years they have become mainstream technologies, whose usage has enabled business users to quickly create data repositories for their analytics needs with little or no IT help.

Hadoop is an open-source software framework that is used to store and process Big Data in a distributed manner using large clusters of cheap hardware. It provides massive data storage and very fast processing, as well as scalability and fault-tolerance. A key benefit is that it significantly reduces the time required to acquire and load diverse datasets. The reason is that data can be loaded quickly, as it does not need to be modeled to fit a relational schema—the data is just dumped into the Hadoop file system as-is. Therefore, many platforms use Hadoop in the landing area to collect data and the staging area for cleaning and integration.

Unlike legacy platforms, Hadoop can store any kind of data in its native format and, also, perform many types of analyses and transformations on that data. It is designed to parallelize data processing across computing nodes to speed computations.

Hadoop has two components: A massively scalable distributed file system, which can support petabytes of data, and the MapReduce engine.

Leveraging of New Technologies

New technologies provide powerful capabilities that the SSA platform can leverage. For example, a NoSQL database can be used to capture Big Data and provide operational intelligence. When integrated with Hadoop, it can create a powerful combination. When such a connection is made using existing APIs, it enables analysts to perform complex queries for

Big Data analysis, while maintaining the efficiency and ease-of-use of a NoSQL database. MPP databases and Hadoop can also be used to provide analytical insight for analysts and data scientists.

The recent development and widespread availability of data preparation and catalog tools has enabled power-users to become self-sufficient to a large degree. They can now prepare their own data without relying on IT. Other tasks that the tools have made possible for them to perform include the ability to search, clean, format, and merge data.

SSA implementations have become easier due to these new technologies, which help meet the needs of the business users. They get what they need, when they need it, and in the way that they need it. On the other hand, IT gets out of the report development task, which suits them.

Impact of Visual Discovery

The ability of users to perform data analysis, and for data architects to perform data modeling, has been made significantly easier by the development of visual discovery tools. Users without knowledge of coding or SQL can easily connect, through point and click, to databases, data warehouses, applications, and files. Subsequently, they can develop reports, interactive dashboards, and applications and, also, perform powerful analysis.

Impact on Data Modeling

In an SSA environment, a significant amount of data modeling is performed by the business users at the business unit level, rather than by IT which used to perform enterprise data modeling in the legacy environments. In the SSA platform, business users pull data from the data repository after which, based on their department or business unit's requirements, they create the desired business views. This causes the data delivery process to speed up because IT is no longer involved.

When IT does get involved, the process slows down considerably because it works with a backlog and, also, needs time to understand the business use case and requirements before it can develop the data model. Another issue is that IT-developed models are difficult to modify because IT operates in a very structured way, which slows the enhancement process. However, business users are more agile and can tweak a model quickly when required.

Chapter 27

PLANNING FOR IMPLEMENTATION

In organizations without an analytics platform, business decisions are frequently made without basing them on any meaningful data analysis. When such organizations start questioning that approach and evaluate options for supporting data-driven decision-making, they need to perform an evaluation and execute some key steps before building the analytics roadmap. The following sections describe the key analyses and steps required for planning and implementing an analytics platform.

Identify the maturity model

There are five levels of self-service maturity as defined by Wayne Eckerson. The first step is to identify the maturity level at which the company is currently operating, which will enable it to determine the roadmap and approach.

In the first level of maturity, users interact with a standalone tool, typically a legacy business intelligence application. In the second level, users interact with spreadsheets and PDFs, which are limited in their ability to drilldown into the data. In the third level of maturity, analytics starts getting embedded directly into the application and the user's workflow. Since users can perform analytics without leaving the application, it enables analysis in context.

In level four, analytics provides additional enhanced benefits by enabling the mining of data, which can provide additional valuable insights, alerts, etc. to the users. This can be shared with others internally or with external customers and partners. The fifth level provides the capability to directly sell data, analytics, or their related capabilities and services.

Build the roadmap

After an analysis of the business intelligence maturity model, an evaluation needs to be performed to determine the current status and, also, where the organization needs to be. The unknowns are the time required to achieve it, cost, as well as what is needed for implementation. The roadmap can include several initiatives for enabling self-service analytics.

To realize the outcome, a number of projects may be needed. The initial roadmap should ensure that it leads to a sustainable analytics platform, which can be extended as needed. When properly done, it can provide justification for a project business case and, hence, an investment that makes financial sense. The overall platform is implemented via a number of projects, with each being implemented after business case validation. None of these projects can individually solve the lack of analytics in the organization. However, having a roadmap helps validate the business case for each project and supports vital investment outcomes.

Align initiatives with the enterprise goals

The value and impact of the self-service analytics initiatives for the organization should be communicated to those impacted by them. The benefits and successes should be linked to the organizational goals, which builds trust and confidence in the approach. When that happens, it makes it easier to provide continued support and, also, encourages more departments and users to get involved.

Adopt Agile

For a self-service analytics program to be successful, just providing access to powerful tools is not a good strategy. It should be accompanied by a change in the culture and how analytics is approached. For this to happen, agile techniques should be used, which will improve the probability of success.

In an agile environment, users will be driven to self-service analytics. When the Agile methodology is used, results are delivered in shorter development cycles which involve multiple iterations. Each iteration leads to the development and rollout into production of a small subset of the overall requirement. If needed, the business requirements can be refined after each iteration. This means that users do not have to wait for a big bang rollout with its associated risks and delays. This leads to better and higher adoption by the users, whose changing business needs are taken into consideration in a timely manner.

Besides an agile approach, the overall approach should be based on smaller projects, with smaller datasets which the users are familiar with. Once the users work on such limited data, it increases confidence and makes them more involved, leading to larger and more challenging initiatives.

Involve the users

Business users should be involved at every stage, from requirements gathering to user acceptance testing, as well as for providing feedback for enhancements and new requirements. In particular, key business users should be involved when the requirements are being specified and documented. This will provide many benefits such as understanding the users, their data and analytics requirements, types of tools to be used, collaboration requirements, scope of training to be provided, etc.

Pick the right tool

The selected self-service analytics tool should meet the requirements of the users as well as IT, which can be different and conflicting in certain aspects. For example, while users favor an easy-to-use tool that is powerful from a reporting and analysis perspective, IT favors one that is easy for administration, maintenance, and governance. Therefore, there have to be compromises between functionality and IT requirements.

The tool to be selected should provide users the ability to perform analysis and reach quality data-driven decisions. However, when the tool's required analytical capabilities are being specified during the requirements phase, they should be matched with the business and technical realities.

The options available to IT are numerous and need to be carefully evaluated. These include cloud-based subscription models, different license costs for different user types, open source, free (so long as some criteria are not exceeded), etc. The evaluation should be comprehensive and consider the vendor's land and expand strategy, which can make the tool appealing for the short-term, as well as the vendor's roadmap for the tool. Buying a tool that removes a pain point but has the potential to create a long-term problem should be avoided.

When the right tools are used to enable self-service, it benefits both the business users as well as IT. It goes a long way to spread the analytics culture and, also, helps shift the focus from IT to the business users. Therefore, the specifications for the analytics tool should be very clear and precise.

Develop onboarding plan

Planning should include the development of a user onboarding process. Once users decide they want to use a platform, they want to get onto it quickly. Therefore, any delay negatively impacts user experience and their desire to use the platform. An onboarding plan involves many elements including role assignment, training, licensing, access/authorization, etc.

Planning should also include a process for onboarding new data onto the platform. Power-users or others with the authorization to work with internal (non-SSA platform) and external data will need to be guided on how to import data and, also, conform to various requirements pertaining to testing, security, and governance. Unless data can be onboarded easily, it will limit the platform's scalability, as well as the users' ability to perform analytics due to the lack of required data.

Chapter 28

IMPLEMENTATION AND ONBOARDING

Start with a pilot project

To ensure the successful implementation of an SSA program, which impacts an organization and leads to major changes, it should be rolled out in phases. It should first be implemented as a pilot project with a small department or group. After a successful pilot or proof-of concept project, the SSA processes can be tweaked based on the lessons learned. After the project is completed, additional changes can be implemented, if needed.

The processes that need to be tested include platform access, training, user roles, user management, licensing, data onboarding, promoting objects across different environments, development, testing, issue management, etc. If the foundational processes and services are validated, fewer issues will be expected in the general rollout and the probability of experiencing major issues will be low.

Develop phased implementation plan

An SSA project can be quite extensive in scope as it can impact thousands of users and involve scores of data sources. Therefore, it is a good idea to implement the project in phases after the initial proof of concept (pilot) project. The phases can be planned based on a timeframe, such as an 18-month program, in which a few new departments or business units are onboarded every month. An alternate plan could be a two or three phase project in which all the departments are onboarded within a specified time period. The plan chosen will depend on the budget, as well as the enterprise and competitive pressure for quickly rolling out analytics.

Install the SSA platform

This is the step in which the SSA infrastructure, architecture components, applications, and tools are installed by the technical team (IT). The installation can include an assortment of hardware and software such as servers, databases, data marts, data lake, Hadoop, front-end tool, as well as platform environments such as development, quality assurance, production and training. A number of software applications will be installed such as monitoring software, ETL software, etc.

Onboard users

The onboarding process will vary from implementation to implementation. The specific steps to be executed will depend on a number of factors. Also, the sequence of steps will be different and, in many cases, some steps will be executed in parallel. The key steps for onboarding a department onto the SSA platform are described in the following sections.

Perform a Preliminary Assessment

An SSA platform may not be appropriate for some department. Making it go through a few gating questions will indicate if there is potential for it to benefit from the platform. This assessment can be performed through a quiz, which can be based on some screening and gating questions.

Fill Out the Questionnaire

If the assessment in the previous step is positive, the department will be asked to provide key information through a questionnaire. The questionnaire can include questions pertaining to the problem to be solved (use case), data sources (number and types), number of users, data classification (as sensitive data will need to be handled differently), data owners, existing content migration requirement, etc.

Interview with the project manager

During the interview, the project manager or the assigned senior technical resource will review the questionnaire. Questions that arise, based on the information provided in the questionnaire, will be clarified. At this stage, it will be determined if there are any special or specific technical features that are required. After the interview process, the level of effort and IT involvement will start becoming clear.

Develop a project plan

In this step, a plan will be developed for onboarding the department's users and data.

Identify the users

In this step, the department users will be identified and they will be assigned appropriate roles. This will determine what they can do once platform access is authorized and, also, which training will be appropriate for them based on their assigned role.

Identify the lead developer

Each department will be asked to designate a lead developer. This resource will be the primary point of contact for all technical requirements for the department to be onboarded. The lead developer, who performs a key role, acts as the intermediary between the department and two key teams: IT and the SSA program team. This resource will coordinate the department's onboarding activities and, also, be responsible for the administration and management of the department's data, reports, dashboards, and users. The lead developer will also be responsible for establishing connections to the data sources and pulling in required data into the platform.

Provide training

In this step, users will take the appropriate training based on their specific role—casual user, power-user, data explorer, or developer. The lead developer will also need additional training on the technical aspects of the platform and tools, including how to manage, maintain, and administer the department's project folders and users.

Provide access to the SSA platform

After the users are trained, they will be setup in the SSA platform with the appropriate access authorization based on their designated roles.

Identify data sources

In this step, the data sources required by the department for building reports and dashboards, as well as for performing analyses, will be identified. These can be a mix of data sources that already exist on the platform, as well as data sources that are not available within the SSA platform and, therefore, will need to be imported.

Onboard data

IT or the lead developer will import the required data into the platform from the data sources that are identified in the previous step.

Start development

Users start working on the platform. The power-users start developing content and the data consumers (end-users) start consuming it.

Provide targeted training

Modern analytics tools are very complex and just providing access to these tools is not a good strategy. Users must be trained in how to use the selected tools for the specific work that they are expected to perform. The level of expertise determines how users are to be trained and exposed to the tools, so that they can use them easily, effectively, securely, and without becoming a support burden for advanced users or IT.

Casual business users can be trained quickly, with minimal training, to learn the basics of navigation, executing reports, and accessing dashboards. On the other hand, users who are more tech-savvy may need more in-depth training, which can include data preparation, data migration, data manipulation and analysis, data modeling, data visualization, data governance, as well as tool-specific advanced training.

Any project plan should include a strategy for providing training to the various types of users expected to use the platform.

Develop training and reference documentation

As part of the program, it is essential that documentation should be developed and provided to the users. The types of documents to be developed can include (though not limited to):

- SSA process documents, which can include SSA process and architecture, as well as processes for onboarding users, platform/front-end tool authorization request, data onboarding, data governance, and training
- Documentation relevant to the data in the SSA platform
- Front-end tool training
- Training documentation by role, such as developer, power-user and end-user
- Training documentation of various types including cheat sheets, best practices, frequently asked questions (FAQs), etc.
- Learning Management System (LMS) training, if it is used for rolling out online training
- Self-Service User Guide, a comprehensive reference document, which provides important information to get an end-user up and running on the SSA platform; covers most topics of interest to the users (such as processes, FAQs, and help)

- Technical Guide, which can include procedures on how to add users, create folders and projects, request new access to an enterprise data source, publish reports and dashboards, publish data sources, migrate objects from environment to environment (development to QA to Production), other technical topics, and best practices

Develop data governance documentation

For data governance, there will be a requirement to develop comprehensive documentation. If a data governance tool like Collibra is available, a fair amount of work can be automated once the SSA platform goes into production. However, when the program is being initially planned and executed, many datasets, reports, dashboards, and metrics will need to be onboarded onto the platform, which can require an extensive manual documentation effort.

The key items for which information will typically be needed by the governance team include applications, data sources, dashboards, reports, metrics, and dimensions. The content for these can include:

- Applications: Data asset type, data source used, sensitive data used (confidential, PHI or PII), data classification level, and owners (business and technical)
- Data sources: Type, applications using the data source, any sensitive data used (confidential, PHI or PII), data classification level, and owners (business and technical)
- Dashboards: Stakeholders, data sources, any sensitive data used (confidential, PHI or PII), data classification level, owners (business and technical), refresh frequency, and access restrictions
- Reports: Stakeholders, owners, data sources, refresh frequency, any sensitive data used (confidential, PHI or PII), and access restrictions
- Metrics and measures: Stakeholders, owners, and definitions
- Other: Business glossary and data dictionary

Chapter 29

TOOLS FOR ANALYTICS

Objective

An analytics tool is a software application that helps users perform reporting and analysis. Its objective is to provide users the ability to analyze data and extract actionable, relevant information which can be used for decision-making. Besides the basic ability to retrieve and analyze data, analytics tools are characterized by powerful capabilities and a wide range of features.

Tool features

Data analytics tools, which are feature rich, are not the best in class in all areas. Therefore, what sets them apart is the ability to excel in a specific area such as data mining, collaboration, data visualization, and predictive or suggestive intelligence. These tools use various types of analysis techniques for analyzing datasets. Many of these tools also generate reports or provide the results through data visualization.

A tool with data mining capability is able to drilldown into the available data and provide analysis through various techniques that search for patterns. A collaboration tool provides platform users the ability to collaborate efficiently and easily with other users across the enterprise. A tool with data visualization capabilities provides a powerful and user-friendly interface, which is one of the key features desired by end-users.

A tool with predictive and suggestive intelligence capabilities is extremely powerful, though it is not one used by casual users. This type of tool, rather than just analyzing and predicting what will happen, goes one step further by providing suggestions for what needs to be done.

Analytics tools are also available in the cloud, which provides some advantages compared to onsite tools. They are scalable and do not need any onsite infrastructure. Typically, they are subscription-based. Their benefits include the ability to provide access from any device or platform, faster connectivity, and better collaboration.

Selection

Data analytics tools are used in organizations of all types and sizes. The prerequisite for selecting such tools, which provide results with minimal effort, is to understand the analytics needs of the users. The challenge is that these users have very different skillsets, as they range from novices to experts who may or may not have a technical background. Also, since the analytics requirements vary across individuals, groups, departments, business units, and the enterprise, the tool required will depend on the specific needs and analysis requirements.

When the needs of the users and the enterprise are clear, it becomes easier to map and select the right tool. If a tool is selected without comprehensive analysis, it might not meet the needs of the users or IT. It may be too simple, too complex, lack critical features, difficult to integrate, or be difficult to administer and support. The key for successful selection is to address the needs of two major groups in the SSA environment, analysts and casual users, who have significantly different requirements. Having either group dissatisfied will cause user adoption to suffer.

Complexity of analytics tools

The complexity of data access tools varies significantly. Desktop-based tools have been supplemented by web-based and cloud-based tools, which provide many benefits. They are more versatile, easy-to-use, eliminate the need to install special software on every desktop, provide access through a simple browser, and reduce costs because they require less powerful and expensive hardware. These tools can be used by users at different ends of the skills spectrum and the corporate hierarchy, as they are quite versatile and able to cater to the needs of diverse user bases.

Variety of analytics tools

A single analytics tool will not meet all the analytics needs of most organizations. Therefore, they will frequently use multiple tools for different groups or types of users, such as power-users and casual users. Any tool that is selected needs to be tailored to the specific skills of the users and the analytics requirements of the organization. For example, the selected tool may need to excel in data mining, working with real-time data, data visualization, advanced analytics, or data preparation.

The key categories of analytic tools are:

- Search: These tools permit users to submit their ad hoc queries by typing key words in a search box.
- BI tools: These tools typically focus on a single aspect of analytics such as visualization, reporting, dashboards, OLAP, etc.
- OLAP: Online analytical processing (OLAP) is a BI tool that addresses the need to perform multi-dimensional analysis. It is based on a technique that combines data access tools with an analytical database engine. In contrast to the rows and columns structure of relational databases, OLAP uses a multi-dimensional view of data (such as sales by brand, season, and store).
- Visual Discovery: These are self-service in-memory analysis tools, initially designed for data analysts, which enable business users to access, analyze, and visualize data. Subsequently, they can share their data discovery results by publishing them onto an interactive dashboard.
- Machine Learning: These tools use algorithms and statistical models that allow software applications to accurately predict outcomes, without being provided explicit programming instructions.

They rely on patterns and inference and, over time, as new data becomes available, update the outputs. These tools are designed for sophisticated data scientists and statisticians.

Best data analytics tools

The appropriateness of a self-service analytics tool depends on many factors, such as the use case, industry, budget, types of users, expertise of available technical resources, BI maturity, and many other variables. The following are currently among the top data analytics tools: R programming, Tableau Public, SAS, Apache Spark, Excel, RapidMiner, KNIME, and QlikView.

Chapter 30

DATA VISUALIZATION

What is data visualization?

Data visualization is the presentation of information and data in graphical format, which enables information to be communicated clearly and efficiently. The human brain processes information faster when it is displayed in a visual format, rather than in spreadsheets or multiple report pages. Therefore, making data visual is a key element of making analytics meaningful to business users.

Drivers

As more data is collected for analysis, decision-makers across enterprises have started using data visualization extensively for presenting data and information because it overcomes the shortcomings of conventional reporting and analysis techniques. Data visualization enables results to be presented visually, determine relevance between various variables, and communicate results to others.

Benefits

Data visualization provides the ability to communicate information clearly and efficiently to users via graphics, such as charts and tables. It can enable analysts to observe previously unknown patterns quickly and easily despite large data volumes. When they are well-designed, data visualizations enable users to analyze and reason about the data and evidence.

They also make complex data more accessible, understandable, and usable. Generally, tables are used when there is a need to look-up a specific measure of a variable, while charts of various types are used to show patterns or relationships in the data for one or more variables.

Data visualization makes it easier for users to work with and analyze data, despite its complexity and large volume. Its business benefits include improved operational efficiency, faster response to business change, ability to identify trends, and higher productivity. All this makes data visualization an important technology and, hence, it is no surprise that it has become an essential component of SSA platforms.

Interactive data visualization

Interactive data visualization goes a step further. It moves beyond the display of static graphics and spreadsheets to drilldown into charts and graphs for more granular information. Users can interactively (and immediately) change the data that is being displayed and, also, how it is processed. For example, a filter or a selection period can be changed on the fly to enable the display of a different business perspective. This can change a static visual into an interactive visualization or dashboard, as needed.

The ability of data visualization to provide interactive and drilldown capability into the underlying data is a very powerful tool in the hands of users. It enables them to determine the next question in their quest for the answer to their original question, such as the cause of declining sales for a particular product.

Visualization types

The most popular data visualization types are bar charts, line charts, and pie charts. Advanced data visualization types include bullet charts, sparkline charts, scatter plots, heat maps, tree maps, and pareto charts.

Best visualizations

Data visualizations should not be designed for displaying artistic talent. Instead, they should be designed with function in mind so that information is conveyed to the users quickly and intuitively. Well-designed visualizations are most powerful and effective when they are simple and able to convey the desired information clearly and directly. Complex and artistic visuals may appear impressive at first glance. However, they usually fail in the primary objective of conveying information to the users quickly and easily without taxing their brains.

Chapter 31
TOOL MAPPING AND SELECTION

Mapping necessity

SSA should be viewed as a platform for increasing the business intelligence base of users and democratizing analytics—not for implementing a technology, tool, or functionality. This requirement creates a complication for SSA projects due to the conflicting demands of the different types of users. For example, senior executives and casual users may have their needs met by interactive reports and dashboards, which they can drilldown into. On the other hand, advanced users may require far more complex functionality and authoring features. Therefore, a key prerequisite for implementing an SSA platform is to map the users to the tools because a cookie-cutter approach will not work. When mapping is done correctly, the unique needs of different user types will be met, leading to satisfied tool users.

How tools are used

It should be assumed that casual users will primarily use analytics tools for displaying reports and also perform simple analysis. Power-users will use the tools for report and dashboard development, connecting to different data sources, data preparation, manipulating datasets, publishing objects that they develop, and also perform ad hoc analyses. Some users, like the data explorers, who sit astride light self-service and full self-service, may use tools which help them analyze data, create or manipulate datasets, or develop ad hoc reports. Data scientists and data analysts may use multiple tools that can include BI tools, data preparation tools, and visual discovery tools.

Mapping users to tools

The mapping exercise will help determine which tools are needed, as well as the number of users to be trained and supported. The exercise will map the user types, such as data consumer, data analyst and data scientist, to:

- Analytic tools (dashboards, BI authoring, visual discovery, and machine learning)
- Data tools (snapshots, embedded data prep—which are natively embedded in analytics platforms, data preparation, SQL, Python, Pig, etc.)

Most traditional business intelligence tools enable their users to create ad hoc content. However, when mapping, it should be realized that every BI and reporting tool is not a visual discovery tool.

Selection variables to be evaluated

For selecting an analytics tool, many variables need to be evaluated. The key considerations are described in the following sections.

Strategic questions

All SSA tools should be carefully evaluated. One of the key considerations is if a tool fits in with the organization's analytics strategic vision. For example, a tool may be ideal for the organization based on its current needs. However, if it will not fit in with the strategic plans or has a potential compatibility issue with a planned infrastructure upgrade or replacement, such as migration to the cloud, alternatives should be considered and evaluated.

Functionality

The tool must have features and functionality that support the mandatory requirements documented in the specifications. The typical desired functionality for a self-service analytics tool include reporting and analysis capabilities, built-in collaboration, sharing, messaging, alerts, in-memory processing, embedded functionality, and integration with other tools (such as a data catalog or document management system). For IT, the requirements include administration capabilities and user management, as well as other important features pertaining to security and governance.

Approach to avoid

A bottom-up approach is better than a top-down approach. The approach taken should focus on the ability of the tool to meet the requirements of the users as well as IT, which maintains and manages the tool. The ability to meet the requirements should be the driver, not a desire to install a new or shiny tool with bells and whistles, which is an approach doomed to failure. Therefore, the users should be analyzed and identified before the tool evaluation takes place. Selecting a more advanced tool than the users need will frustrate them and lead to the development of shadow IT groups, along with their associated problems.

Ask the right questions

It is imperative that tool vendors should be asked the right questions, which becomes possible by determining the:

- Business questions that the users are trying to answer
- Familiarity of users with analytics
- Maturity level the organization wants to achieve in the next few years

Some of the questions for which answers should be sought include must have features and functions, ability to meet the requirements of the diverse platform user population, vendor support expected for the platform post-installation, support through user community, scalability, deployment model (desktop, web, and cloud), etc. Getting the right answers will help select a tool of the type and scale that can meet the requirements or, alternatively, lead to a decision to use multiple tools.

Desired features of front-end tools

The following are the most common features desired by the vast majority of analytics users:

- Different types of visualizations such as lines charts, bar charts, gauges, bullet charts, heat maps, tree maps, sparklines, scatterplots, trendlines, and pie charts
- Filtering which can be performed through single value or multiple value selections, sliders, radio buttons, checkboxes, and dropdowns
- Analysis through drilldown, drill across, swapping, crosstabs, grouping, sorting, aggregation, trending, forecasting and hierarchies
- Print, download data, and export results to Excel, PDF or PowerPoint
- Add, delete, and rename dimensions and measures
- Formatting
- Collaboration
- Create dashboards and stories

Mobile devices

The ability to view reports and dashboards and, also, perform analysis on mobile devices is important for executives and employees on the go. Having information at their fingertips on mobile devices is a requirement that has evolved from "nice to have." It is now a mandatory requirement for many enterprises, who want their employees to have access to information on different device types, when they want it, and where they want it. Therefore, the ability of a tool to present the results in an appealing way onto multiple devices, with minimal or no customization, is an important evaluation criterion.

Chapter 32

DATA GOVERNANCE

What is data governance?

Data governance is a data management concept that refers to the strategy of managing and controlling data, which requires process, technology, and people. It covers the overall management of the integrity, consistency, security, usability, and availability of an organization's data.

Data governance is implemented using processes and practices that ensure the monitoring and enforcement of controls pertaining to the organization's data assets, as well as accountability. It is executed based on processes that specify what action can be taken, by whom, and when. It also specifies the conditions under which these actions can be undertaken, as well as the methods to be used.

Poor data governance in SSA platforms

The impetus for the development of self-service analytics capabilities came from the limitations of the traditional BI systems. The SSA approach tried to overcome two serious legacy BI limitations: reporting backlog and limited agility. When SSA became available, business users started using self-service tools at an unprecedented rate, which met their immediate needs but created long-term problems. While the SSA solution provided freedom to users, it was not complemented by robust governance policies and checks, which led to many data-related issues.

Why data governance is needed

If there is no governance or it is weak, it will provide limited visibility into the data. For example, users will be unable to trace data lineage—what changes were made, who made the changes, and when they were made. Also, interpreting the available data may be difficult or simply not possible. Other common issues include data integrity, data and analysis inconsistency, unauthorized access and data security, inability to provide a single version of the truth, as well as difficulty in finding data required to perform the desired analyses.

Problems due to lack of governance

When data governance is not enforced, complications that can occur include poor data quality, data integration, and logical errors. Poor data quality is a common issue for self-service analytics, which can be due to inconsistent definitions. For example, different departments may have different definitions for the same dimension or measure, such as net sales. To avoid inconsistency, every new metric introduced into the platform must be checked against existing metrics. If any discrepancies are noted, they should be addressed before being approved.

Data integration issues can occur when data is not correctly formatted prior to it being aggregated or shared, which will lead to incorrect results. This issue can be reduced, to some extent, through the use of relevant tools and templates. Logical errors can be attributed to using the wrong dataset, faulty merged data, or faulty calculations. These issues can be reduced by implementing a validation process, which will prevent such data from being used for some actions or decision-making.

Data governance strategy

A self-service analytics platform cannot function properly unless there exists a data governance strategy. A successful self-service strategy will incorporate data governance because it will ensure that the information being shared is accurate and of high quality. Data is a corporate asset and should be treated as such. Just making IT responsible will not ensure data integrity, safety, and governance. It has to be done by incorporating a data governance program in partnership with the business.

Data governance requirements

To produce and manage governed data, the key requirements include developing a data management strategy, creating an enterprise structure for data management, identifying and defining data to be governed, developing appropriate processes and, also, publishing data for the benefit of the enterprise users. Other tasks associated with implementing data governance include developing a data catalog, services for processing data, rolling out a collaboration process, installing a data management tool, as well as implementing security requirements.

In an SSA environment, data governance should be implemented with features such as data lineage display, version control, role-based data access, ability to migrate or export data to a different repository, collaboration, administrative features, etc.

Role of governance

Data governance has an important role to play in an SSA environment, where IT is actively involved in its implementation and monitoring. IT has many responsibilities that affect the way governance is implemented pertaining to data access, data quality, mapping of data relationships, etc.

Most organizations implement some form of governance in order to protect sensitive data from getting into unauthorized hands. Depending on the industry, regulatory and compliance requirements, and the level of acceptable risk, the level of governance that is implemented by an organization can vary. The most stringent governance is implemented in healthcare and banking.

Benefits of data governance

Data governance provides many benefits such as data integrity, security, reliability, and data quality. It prevents data redundancy, ensures that the same metric definitions are used across the organization, provides higher information accuracy, enables transparency, provides ability to trace data lineage, forces standardization, increases value of data, creates trust in the data and analyses, increases operational efficiency, enables better collaboration across groups and departments, improves analytics and decision-making, ensures regulatory and compliance requirements are met, and provides better control and management of data assets.

Data governance provides trusted, consistent, and reliable data across the enterprise in a unified format, which enables insight that would not be possible in a non-governed system. By providing visibility and enabling users to understand the platform data, they can make better informed decisions and, also, select the right data or model to use or explore.

Data governance enables a collaborative framework that can be used to define and manage organization-wide rules and data assets. Its reach extends to data development, data security management, data architecture management, data quality management, business intelligence management, as well as document and content management.

Data governance provides control and rule-based usage restrictions. It permits controlled flexibility and freedom without compromising on data quality and security. It also provides the best way to maintain a single version of the truth.

Data governance framework

A data governance framework is a structure for building a model for managing enterprise data. It provides guidelines and rules for complex activities, which deal with or lead to data creation or manipulation. A data governance framework is a critical requirement for any self-service environment, which should include:

- Metadata management
- Master data management
- Data quality management

For SSA to be successful, the business and IT need to be well aligned. They need to be fully aware of their roles and responsibilities, take ownership of their areas of responsibility, and be fully engaged. The data steward, who plays an important role in the governance process, acts as the liaison between IT and the business. Due to their central role, data stewards have to work with and establish a smooth working relationship with IT, SMEs, developers, and power-users, as well as other data stewards.

Implementing governed SSA

Responsibility

A data governance council with responsibility for data governance across the enterprise, throughout the analytics lifecycle, should be setup at the enterprise level. This group should develop the data governance strategy and approve the data management process, policies, rules, and restrictions. It should provide guidance on data quality, other data-related topics and, also, identify datasets that can be made available to various groups.

Approach

In an ideal enterprise environment, only officially-approved and installed BI tools can be used to access enterprise data. In such an environment, IT will not be resource constrained, which will make them super responsive to the needs of the users and, hence, there will be no backlog and dissatisfied users. Such a scenario will ensure a governed self-service environment managed by IT, which will be responsible for managing the platform, data, reports, dashboards, access, and users. It will also manage collaboration. However, this is not a realistic scenario in any organization.

In the practical world, an SSA platform starts small and, over time, expands its footprint. As more data becomes available, it is brought into the platform and incorporated into the enterprise data model. Over time, most or all of the data needed by the enterprise users is onboarded, certified by the governance team, and then made available to the users.

Risk due to analytics tools

The implementation of effective governance has been hampered by the lack of controls in the most commonly used analytics tools. If effective platform controls are not in place, providing business users with analytics tools can become a serious security and privacy risk. However, in a controlled governed system, this issue is avoided. If a user tries to share data or objects such as reports and dashboards with others, controls implemented by IT will prevent such an action.

Enforcing governance

Publishing data governance policies and rules will not be sufficient in an SSA environment. Even though a majority of users may follow them, even a single rogue user can create major issues. Therefore, it is imperative that controls for enforcing data governance should be implemented. Implementing reasonable checks provides a necessary limitation on user autonomy, while ensuring proper governance as well as the benefits of self-service analytics.

Chapter 33

GOVERNED SELF-SERVICE ANALYTICS

In order to implement governed self-service analytics across an organization, the following essential tasks need to be completed:

- Install
- Monitor
- Certify
- Promote

Install

The first step in implementing an SSA platform is to prepare the SSA infrastructure and tools, which is the responsibility of IT. The next step is to provide the developers and business users with the appropriate privileges and authorizations, which will enable them to access the platform, connect to data sources, and import data. After the data becomes available, authorized users can develop reports and dashboards and, also, share content.

During installation, configuration occurs at a very granular level, typically by role, which determines what exactly a user can do. For example, a user may be permitted to develop a report or dashboard but restricted from connecting to an existing data source or adding a new data source.

By default, most features are restricted and have to be turned on, either during the installation phase or later on when the platform is operating.

Monitor

After various platform components have been installed, they need to be monitored by the platform administrator. This task is performed using various monitoring tools. The monitoring can be extensive and encompass platform usage from different perspectives, users, content (reports and dashboards), data sources, platform performance, tools, operation of the infrastructure components (such as servers and databases), ETL jobs, migration of objects across environments (development through production), as well as governance and security.

Certify

All content in an SSA environment must be restricted by default and should be monitored from different perspectives including usage, performance, security, governance, and compliance. Any new content developed within the platform, as well as content imported into the environment (such as a new data source), must be reviewed to ensure that it meets the requirements and standards pertaining to governance, security, and performance. An object is certified and introduced into the governed environment only after it passes specified criteria.

When a data source is brought into the development environment, it undergoes light screening. When it is moved into the quality assurance environment, it undergoes additional scrutiny. Finally, when it is to be moved into the production environment, it is exposed to rigorous screening to ensure that it meets all the requirements for migration into production. As part of the analysis, IT and the governance teams have to ensure that the new data source is not a duplicate of data that already exists within the platform. This can be verified through the data catalog, which contains the inventory and lineage of all available data.

Promote

In a well-designed SSA environment, there are three environments:

- Development
- Quality Assurance (QA)
- Production

An object such as a report or dashboard is first developed in the development environment, which is like a sandbox. After the developer is satisfied, it is promoted (migrated) to the QA environment. In a restrictive organization, promotion from the development environment to the QA environment is performed by IT.

In the QA environment, business users perform testing and sign-off. At this stage, the governance team performs its review and provides approval, after which the dataset, metric, report, or dashboard can be moved into production. The migration to production is performed by IT, which also implements the required security and access controls. In some cases, promotion into production can be performed by the developers or power-users, if they are highly skilled and authorized by the IT team.

Chapter 34

MORE ABOUT GOVERNANCE

Risks

In an SSA environment, with its flexibility and empowerment of end-users, the potential for breaking governance rules is greater than in a closed traditional business intelligence system with strict controls managed by IT. SSA users can bring in external non-validated data, provide access to unauthorized users, introduce defective calculated fields or metrics, and also perform other actions not conforming to governance or security rules.

Creating balance

An SSA environment requires balancing governance and agility, which is a challenge for all organizations. To make the system a success, there needs to be an application of the right amount of governance for different architecture components. Loose governance creates data chaos, leading to conflicting information being generated and shared. Too rigid governance limits access to data and, consequently, the ability to perform meaningful analysis which, ultimately, turns away users. When governance controls are balanced, the platform will be safe and yet provide users the freedom to access and analyze data flexibly and quickly, without being dependent on IT except in a limited way.

While the platform goal is to empower users, their ability to create chaos should be limited. This requires collaboration between IT, which is responsible for the enterprise data, and the power-users. The balance reached will be at different levels in different organizations because they are unique businesses and, also, their SSA implementations will never be identical. Therefore, every implementation must be tailored for the specific organization, as well as the users who will access its SSA platform.

Data steward responsibilities

Data stewards have a key role in an SSA environment. They have many responsibilities including creating and documenting data definitions, identifying the relationships between objects, establishing and documenting the context of the content (such as data lineage and heritage), and managing metadata. In addition to maintaining the integrity of the data elements, content, and metadata, they also certify their accuracy and completeness. Data steward responsibilities also include overseeing the lifecycle of a particular dataset, administering the processes, policies, and guidelines applicable to the organization's data, as well as ensuring that all regulatory, legal, and compliance requirements are met.

Data stewards are also responsible for defining and implementing policies and procedures pertaining to the routine operation and administration of systems and data, managing the data glossary and data dictionary, ensuring that all data elements in the environment are defined without ambiguity, providing users with a contextual understanding of the data, as well as monitoring data import, storage, processing, and transmission for all systems (internal or external).

Other data steward responsibilities

Data stewards, who are responsible for data quality, specify the data quality metrics and requirements, as well as the acceptable values and ranges for each data element. They establish the processes and controls for maintaining data quality, as well as for detecting and rectifying data quality anomalies and problems.

Data steward responsibilities also extend to data protection, which is a challenge due to fast data proliferation. They establish the processes and controls that ensure privacy and security pertaining to the datasets they are responsible for. In conformity with the organization's requirements, they have to define policies pertaining to retention, archival, and disposal. Data stewards also define the procedures for accessing data, based on the user role. They work closely with IT to establish the required controls.

Data governance tools

Data governance tools are used to implement and automate the data governance processes and also ensure data integrity. A data governance tool performs traditional data management functions, which address different areas such as master data management (MDM) and security.

The key areas that data governance tools cover include enterprise governance, data lifecycle, metadata management, data remediation, data dictionary, validation of business rules, security, auditing, and reporting, as well as integration with other systems and applications.

Concerns regarding under-the-radar groups

Business users concerns

When shadow IT groups develop, silos are developed. Systems developed by these groups are not well-architected, do not perform optimally, use many poorly modeled independent data structures like data marts and spreadsheets, and are not governed. These systems are not easy to maintain and therefore, when technical issues develop, the business users have to reach out to IT for support. The result is that both IT and the business have to spend time on issues, which pull them away from their primary tasks.

Since only limited local data is available to shadow IT groups, the breadth of analysis that they can perform is limited as well. They cannot access or analyze enterprise data or data owned by other departments because it is not available to them. Another problem is that the results produced by one shadow group may be challenged by others. In many cases, the results are out of sync with the analysis produced by others, due to different definitions and KPIs being used.

IT organization concerns

A nightmare for IT is the need to provide support for shadow systems, which is inefficient and takes away its resources from more productive tasks. Validating or reconciling data, reports, or dashboards produced by shadow IT groups is a problem for IT. Another concern is governance, which cannot be managed easily when there are numerous shadow groups in operation, many of whom are under the radar.

Long-term strategic concerns

An organization having multiple data infrastructures makes inefficient use of its data assets. If left unchecked, there will be data chaos and the platform will have to face serious issues like trust, reliability, and inability to provide analyses that the organization needs. It will also have a negative impact on the organization's analytical maturity. Hence, it is in the long-term strategic interest to bring all groups under a single umbrella, where they can be subjected to governance rules and requirements.

Delays due to governance checks

A member from the governance team or the data steward performs the first check of a newly developed report by comparing it against the existing inventory of reports. The check reviews for duplication of reports. The dimensions and measures in the report are cross-referenced, against those that already exist, to ensure that the definitions are accurate. This quality check can slow the deployment of the report into production and, therefore, cause frustration. However, this must be done to maintain the integrity and quality of the SSA environment and, also, to avoid the extra work that will be required if errors are identified at a later stage.

Chapter 35

DATA CATALOG

What is a data catalog?

A data catalog is a repository that contains metadata. It includes base tables, views, synonyms, and indexes. A data catalog centralizes all information about the data in one central location, which means it contains its structure, definitions, and other relevant information. It helps users understand their data and its context. Through metadata, they can understand the data elements including their origin, time stamp, lineage, owner, etc. This enables data users to search, discover, and consume data sources for their development or analysis.

Need for a data catalog

In an SSA environment data quality is of extremely importance. Unless data is clean and reliable, it cannot be used with any confidence. To ensure data quality, a self-service analytics architecture must include a robust and comprehensive data catalog, which is the central source of truth.

A data catalog is used to properly organize, integrate, and curate data for the users. Over time, it includes the growing list of data sources added to the SSA platform. It enables users to determine which data is available as well as its storage location, which could be a data warehouse, an SQL database, a data lake, an Excel spreadsheet, or some other type of storage structure. Due to its central location, the catalog becomes the single source of truth repository, which contains data that can be used with confidence. It helps avoid redundancy and potential costly mistakes due to the incorrect use of data, provides a simplified view of data, and organizes data across existing silos and boundaries (of systems, business units, and geographies).

Through the data catalog, users can determine what the data means, its origin, how to access it, its accuracy, and how complete it is. When a trusted data catalog is available, users can focus on how to perform analysis, rather than worry about the accuracy of the data.

Data catalog capabilities

For a data catalog to be effective and useful, it should be user-friendly, as its users will not be limited to tech-savvy professionals. Therefore, its navigation should be intuitive, support drag and drop, and also provide intuitive mapping. A user should be able to easily search the catalog and, also, check the data lineage and relationships.

A data catalog should have a flexible structure that can link data sources, applications, and metadata sources to create a system that aligns data to the business. This will enable any modifications to be detected immediately, which can then trigger the appropriate policies to be applied without any manual intervention. A user can develop an analytics model and easily analyze the associated data and its properties, such as its source or history, without needing to verify its authenticity because it is part of the governed data.

In an analytics environment, collaboration is important, as it improves productivity. A data catalog can help integrate the work of multiple users working independently. For example, a power-user needing to build a report can use a report already developed by another platform user, as-is or with some modifications. This sharing can be enabled through tagging and annotations in the catalog. Another way in which a data catalog can be made more useful is by incorporating machine learning functionality, through semantic search capabilities. When this is done, data searches can become more efficient over time.

Key for success

A data catalog can only be successful if it is trusted by the users. Therefore, it should be tightly controlled and managed so that its authenticity is never challenged. If questions are raised about its authenticity, it will lose its greatest benefit of being the trusted source of data. So long as the trust is there, business users will use it for obtaining business insight and decision-making.

Chapter 36

SECURITY

Need for security

In the self-service analytics environment, the importance of security cannot be minimized. All data must be secure and privacy must be ensured. Only those with a need to know should have access to the platform or any data, metric, report, and dashboard. If security is not maintained, especially in highly regulated environments such as healthcare and financial services, the potential for financial losses can be significant in addition to legal, compliance, and regulatory penalties. Despite these well-known issues and risks, some self-service platforms do not have proper and formal data security and data governance controls.

To ensure that the platform as well as its data are secure, the security team should be actively involved in performing security checks before and after implementation. This should not be a one-time exercise and, instead, there should be an ongoing operation to ensure that the security and governance rules are adhered to.

Data privacy

Data can be very sensitive and needs to be protected with extra care for some types of data, such as personal identifiable information (PII) and protected health information (PHI). Data privacy governs how data can be collected, stored, shared, and used.

When working with data, users need to be aware of how they should handle consent, notice, sensitivity, and regulatory requirements. Therefore, rules pertaining to data should be clearly specified so that the platform users are well informed. They should know:

- If or how data can be shared internally or externally with third-parties
- If data can be collected or stored legally
- If there are restrictions due to regulations like GDPR and HIPPA

Platform permissions

The analytics tool vendors are aware of the need to balance self-service with governance. Therefore, many have developed tools that support the implementation of permissions at a very granular level. Permissions can be based on the existing enterprise security model or architecture, such as LDAP and Active Directory. In such cases, administrators do not have to maintain a separate security model. Instead, they can assign permissions and access authorizations, for the platform or individual objects like reports, based on the existing enterprise security model.

Permissions, which can be set at a very granular level, can be different for the various environments (development, QA and production). For example, a user may be granted permission to develop reports in the QA environment. However, in the production environment, such a user may only be allowed to view reports.

Permissions can be based on who can:

- View or save an object
- Save an object to the local drive or onto the server
- Edit or copy existing reports and dashboards
- Develop new reports and dashboards
- Publish a report or a dashboard
- Share a report
- Download data (partial or full data associated with a report)
- Schedule a report for execution or distribution

- Add a new data source or modify an existing data source
- Connect to a new or existing data source
- Develop a new report using ungoverned data (for the appropriate environment)
- Add new dimensions and metrics
- Create calculated fields

Sharing

An important feature desired in an SSA environment is the ability of users to collaborate and share the content they develop with other users. While sharing is a very useful feature, it also entails risks as content, restricted or confidential, may be shared with users who are not authorized to view it. Therefore, when setting up the sharing capability, security should be implemented, though not so rigidly that it becomes an unnecessary barrier.

Balancing risks with ease of use

All SSA platforms need to balance risks with ease of use, which are two requirements that pull in different directions. A very secure environment can be provided by minimizing access and implementing controls, which can become a barrier for ease of use. Providing easy access can be implemented by relaxing security requirements, which increases the risk. Therefore, there needs to be a balance between these two conflicting requirements so that both sides, business users and IT, are satisfied.

Security planning

For enabling and managing self-service, security needs to be well planned and implemented—it cannot be an afterthought. Organizations that start evaluating security right from the start are better able to assess options and make calculated choices, based on their business environment and information sensitivity. The threats over time may change, but the fundamentals to be considered remain the same.

Risk of assumptions

Making assumptions about security can be a big mistake. Trust is fine but it must be based on processes and tools that have been validated by the experts. It should not be assumed that an SSA platform has proper governance, security, and controls. They should be verified before being trusted.

Types of threats

There are many types of threats that an SSA environment can face: external as well as internal. Adequate protection needs to be implemented against both types of threats. Threats can be due to rogue software, rootkit, hacking, viruses, spyware, malware, phishing, DNS spoofing, IP spoofing, SSL hijacking, wi-fi hacking, etc. Internal security failures can be due to unauthorized access to data, misuse by authorized as well as unauthorized staff, privilege misuse, password sharing, data leaks, social engineering, downloading of malicious Internet content, illegal activities, and user errors.

Causes of breaches

The most common reason for security failures can be traced to privilege misuse and human error. These are more likely to happen due to convenience, rather than malicious intent. Even if the infrastructure has good security, it can be negated by an employee's inaction or action performed due to convenience, without regard for its security implications. One estimate is that 85% of threats originate internally.

Need for being proactive

To ensure the safety of the SSA platform and to limit the harm that might be inflicted, in case there are any failures, periodic audits must be conducted. These should be supplemented by random audits.

Chapter 37

IMPLEMENTING SECURITY

Basic steps

The key basic steps that need to be taken to secure data on the platform include:

- Review the security architecture, tools, and policies
- Establish what the desired security state should be
- Identify gaps
- Identify data that the platform contains and where it resides
- Retain data that is required for business operations
- Protect the identified data
- Dispose of data that is not needed
- Prioritize data based on severity and ease of remediation, as importance of data assets varies considerably
- Perform risk assessment which should include identifying threats, determining risk and impact, and analysis of the control environment
- Develop a data security plan

Implementation steps

There are four basic steps involved in data security implementation. The first step is to understand the data and locate where the data resides, especially the sensitive data. This data should be classified after which policies should be defined that explicitly explain who can access the data. The policy should also explain the conditions under which this data can be accessed and managed.

The next step requires the data to be secured and protected. Authorizations are implemented so that only approved users can access the data for which they have been authorized. Sensitive data, such as PHI or PII, will need to be encrypted, de-identified, masked, or redacted, as needed. Procedures will need to be developed, along with appropriate authorizations, for providing users the ability to view or download data (all or partial).

After the data has been secured, it needs to be monitored and audited. This will ensure compliance and enforcement of the security policy. To confirm compliance, automated customized reporting can be developed.

Ensuring control

Users are the biggest security risk for an organization for many reasons, especially due to abuse of privilege access, knowingly or unknowingly. Having policies that ensure controls are in place will prevent security issues and reduce risk. Having a default policy of providing the lowest level of access, which is sufficient for users to perform their work, should be implemented as it will limit risk.

Develop response plan

Even with the best security plan, data breaches will occur. Therefore, it is necessary to have a plan in place that will be initiated if there is a data breach. The logical steps in such a plan are to identify, contain, eliminate, recover, and analyze. This should be followed by developing a plan to address the next attack.

Assign responsibility

Data security responsibility must be assigned to a specific team, which should be tasked with implementing the data security strategy. A security strategy can only be successful if responsibilities are assigned to specific individuals, who will lead the effort to identify, contain, and eliminate a breach when it occurs. This will ensure that everyone involved knows what is expected of them. As needed, these individuals must be provided relevant training so that they can execute their responsibilities effectively.

Supplement IT security staff

In the case where available security and governance expertise is not available, third-party experts can be used to supplement the in-house resources. These partners can be used indefinitely or till the internal staff is adequately trained and ready to take over the responsibilities.

Best practices

Best practices that should be followed to minimize data security breaches. These include:

- Use strong passwords, which should be changed frequently
- Install anti-virus and anti-malware software
- Install a strong firewall, which will protect the network by controlling Internet traffic flow
- Regularly update programs, which will bring their security level up-to-date
- Secure devices, such as laptops and smartphones
- Never become complacent and stop monitoring
- Be aware of risks associated with e-mail, the Internet, and messaging

Training

The best defense is having an informed user community. Users must be provided appropriate data security training. Once they are aware of the risks and what constitutes risky or dangerous behavior, most users will act in a responsible way that will help prevent data breaches.

Chapter 38

TRAINING

Need for training

Modern self-service analytics tools, with a user-friendly interface and point and click capabilities, have a reputation for ease of use and smooth ramp up even for novice users. However, in reality, deployment is not as easy as it appears to be. Users need to be trained on diverse business and technical topics. The specific training will depend on the user role, individual skills, and the specific needs of the organization.

User skills

The technical, analytical, and business skill levels and proficiency vary tremendously within organizations. These range from resources with limited technical skills to highly technical professionals, such as developers, data scientists, and power-users. Therefore, the training of these diverse users will need to be customized, according to their specific needs and the roles to which they are assigned.

Who needs to be trained?

Every user who interacts with an SSA platform needs training of some type. The extent and depth of training will vary according to the specific needs of the user, department, enterprise, or implementation project.

Training curriculum

The training to be provided to the SSA platform users will typically include:

- Processes
- Role-based tool training
- Data visualization
- Data (analysis, modeling, preparation, merging, classification, etc.)
- Developer training
- Tool administration
- Data governance
- Security
- User administration
- Learning Management System (for online training)
- Best practices

These topics can be covered through multiple courses, which should be developed as part of a curriculum for the SSA program. The specific courses for an SSA platform user will depend on the role, existing technical and business skills, familiarity with analytics and the platform, as well as compliance and regulatory requirements.

Training delivery methods

Many options exist for delivering training to the platform users. These include classroom-based instructor-led training (ILT), online training (instructor-led and self-paced), hybrid training, videos, and webinars. The training documentation can be developed internally or, alternatively, courses developed by the platform vendors can be used. For a very large user base, especially for the casual users, the train-the-trainer model for delivering training can be used. For end-user training, power-users should be leveraged to the maximum. Power-users, besides knowing the tool, also understand the business very well and, therefore, are in a better position to address questions and concerns raised by the casual users.

IT training role

IT has an important role in driving user adoption of an SSA platform. They can help train power-users with data and analysis related topics, as well as security and governance. If these users become proficient, they can produce content that will drive users to the platform.

IT also has an important role in onboarding early adopters, as it can help them become familiar with the environment, tools, data modeling, data onboarding, as well as the process for promoting content from development to production. IT can also teach them best practices, which will help produce better content and, also, avoid development that becomes a performance nightmare.

Chapter 39

BARRIERS TO SSA ADOPTION

Despite the benefits associated with SSA, there are many barriers that need to be overcome to make it a success. Just providing access to the platform is not sufficient. Business users need to be trained, provided appropriate access, and empowered so that they can produce their own reports, dashboards, and analyses. The barriers that can come in the way of self-service adoption are described in the following sections.

Less than expected usage

In many organizations, the desired level of participation is less than one would expect in a self-service analytics platform, which can usually be attributed to dissatisfied users. The key reasons for the less than desired participation include:

- Lack of interest: Not everyone wants to be self-sufficient and many users/managers prefer that the results be presented to them in the final form
- Data preparation: Cleaning, transforming, and blending data is a barrier for users with limited technical skills
- Access to data: Only experienced users can independently access and query Hadoop or other Big Data platforms
- Limited budget: Tools to access and analyze data in a self-service analytics platform can be expensive; when there are thousands of

users, the total cost of licenses can be very high and, therefore, they are rationed which limits participation

Misplaced expectations

SSA adoption rates are still low for a number of reasons, including the perception of many users that it is too slow. Typical users have a good idea of what they want but they don't want to wait for it. Such users expect to be up and running very quickly and want to produce reports, dashboards, and analyses at lightning speed. In the world of instant search, these expectations are misplaced.

Another reason for the frustration is the time it takes to prepare a data source, which may involve modifying an existing data source, adding a new source, adding or modifying a new data connection, etc. These potential issues can be addressed through appropriate communication and management of expectations.

Complexity

Getting users to use self-service analytics is not always as simple as it appears to be. A leading reason for low adoption is that users find the analytics tool complex and difficult to apply to their business problems. Instead of simplifying analytics, tool complexity makes life difficult for users, especially if they are not provided adequate and appropriate training. When expectations of ease of use and independence are not met, frustration rises and users stay away from the tool and platform, unless absolutely necessary or mandated.

Functionality limited by default

Self-service analytics tools are very powerful and can do a lot of damage in the wrong hands. Therefore, on many platforms, the default permissions for users limits the features that they can access or execute. For example, features that are suppressed include filtering, sharing, saving, downloading results, developing dashboards, and publishing workbooks to the server.

Those who desire more features than they are provided by default can initiate a request for those capabilities. Since there are so many SSA platform users, a process needs to be designed, ideally through workflow, so that such requests are handled in an expedited way. Lack of such a process can become a barrier to adoption across the enterprise.

Rigid data governance

Implementing sound data governance has many benefits. However, when it is very rigid and inflexible, it becomes a barrier for the users. When that happens, it negatively impacts adoption across the organization. The governance process should strike a balance between flexibility and security so that it meets the needs of governance, IT, as well as the business users.

Project is tool-driven

A self-service analytics project should be process-driven, rather than tool driven. When an SSA program is planned and implemented around a preselected tool, its chances of success are low compared to a project that is process-driven. If a tool is selected first, it may or may not be the appropriate one for the types of users who will be using it.

Analyzing users and how they perform reporting and analysis will highlight the appropriate tool features and capabilities. Based on such analysis, the appropriate tool can be selected and procured. When a tool is preselected and the SSA users find it limited or unsuitable for the work that they perform, it will hinder usage and adoption.

Single tool strategy

Since a long time, enterprises have been trying to minimize the number of tools for their business intelligence needs in order to standardize, deal with fewer vendors, and reduce costs. However, restricting users to a single front-end tool creates a challenge.

An SSA environment is characterized by thousands of users who have different roles: end-user, power-user, analyst, developer, and data scientist. Since the requirements of these roles vary significantly, it cannot be expected that all users will use the same features and interact with the software in the same way. When a tool is very complex, it can meet the requirements of the analysts. However, a casual user might find that tool too complex. The reverse also holds true. Therefore, a one-size-fits-all tool strategy is not the right one if high self-service adoption rates are desired and, therefore, enterprises should consider providing multiple tools.

Limited tool licenses

As an SSA program starts getting more popular, more users clamor to get onto the platform. This creates a big demand for front-end tool licenses, which can be expensive. If an enterprise has a licensing model which limits the number of licenses that can be assigned, they will need to be rationed among departments and users. This will cause fewer users to use the platform compared to the scenario where licenses are issued liberally.

When it exists, the license allocation policy creates many unhappy users who are dissatisfied with the license distribution process. Therefore, in such a case, a clearly defined license issue policy should be communicated and implemented.

Lack of data for required analysis

Self-service analytics is based on data that has been made available to the platform, which will typically not be all of the enterprise data, which is stored in different systems. While the available data can support useful analyses, it may not be sufficient for comprehensive enterprise-wide analyses. If some key data is missing, or the user does not have access to it due to lack of authorization, it will prevent meaningful analysis in such a case. When data accessibility and visibility are limited or prevented, it can become a barrier to adoption for the impacted users.

Poor data accuracy

If there is any issue with data accuracy, which makes users lose trust in the data, the program will end up being a failure as its output will not be trusted. Some of the common issues leading to poor data accuracy include duplicated values, missing fields, corrupt data, incorrect mapping, corruption due to ETL issues, stale data due to collection lag, etc.

Lack of workflow integration

When self-service analytics is integrated with a business application in a user's daily workflow, it significantly improves the overall efficiency and performance. Such a user starts to better understand the data and its impact, explores the data to gain insights previously not possible, makes decisions that are data-driven, becomes more inquisitive, and starts to analyze the data from different perspectives. User adoption will improve if workflow integration becomes widely used, which is not the case at this time.

Poor relationship with IT

A self-service analytics program will be successful if there is good partnership between IT and the business. Their responsibilities should be balanced and IT should try to provide maximum flexibility to the business users. When IT implements controls rigidly, which impacts the ability of users to work effectively, they will stay away from the platform which, ultimately, will lead to lower adoption. Therefore, it is very important to have a good working relationship between the business and IT. When their partnership does not exist, the platform will be unsuccessful in achieving its objectives and attracting users.

Lack of training

Getting an SSA platform infrastructure up and running is not sufficient for a rollout. The platform users need to be prepared for the new way of working with data and performing analysis. They also need to get familiar with a tool, which they probably have never used previously. The SSA program must prepare its users with customized role-based training. Depending on the audience, the training can include process, tools, data, and governance.

When training is not provided, it can become a serious barrier for adoption. Be aware that many users will need a lot of hand-holding, which may not seem obvious in a self-service analytics environment. Training support will need to be provided on an ongoing basis, as it will be required for new users and for those who want to learn advanced tool features. If support is not provided or is inadequate, it will inhibit platform adoption.

Chapter 40

IMPROVING ADOPTION

Reason for steady adoption rates

The adoption rate of self-service analytics has remained fairly steady over the past few years. This is somewhat surprising considering the potential benefits that can accrue through its implementation. An explanation provided for this is that some self-service systems may not be classified as such. However, since most business intelligence tools provide some self-service capabilities, the actual number may be higher.

Focus on the users

Building the platform and making a reporting and analysis tool available to the users is only the first step. This should be accompanied by steps that avoid or remove adoption roadblocks and, also, enable employees to use the tool. To ensure success, an SSA program should empower users so that they are not dependent for their routine work on IT and power-users, who are in short supply. The program should include a change management plan in order to soften the disruption caused by introducing SSA. A support process should also be developed, along with the establishment of a Center of Excellence.

Align requirements with user needs

The business requirements for an SSA program should be top-driven. The program should be led by someone who has a vested stake in the project outcome. The business requirements and the project scope should be clearly defined if failure is to be avoided. These requirements, which can be flushed out in workshops, should be well-aligned with the needs of the users and the business. Ideally, the requirements should be modest in the initial phase to reduce risk. After experience has been gained, the program can be rolled out across the enterprise.

Start small

The SSA program should not be rolled out with a big bang. The initial phase of the project should be limited to a few departments and groups. After they have been identified, select 1-2 groups for the initial proof-of-concept (POC) pilot implementation. The pilot project will enable testing of the processes and tools, which will provide valuable feedback that can be used to tweak the processes and training. If there is any issue that arises during the POC, it will be limited to only a few users.

Manage expectations

Expectations need to be managed to ensure that there are no disappointments. A message to share is that while the self-service analytics program will provide many benefits, especially flexibility and a shorter development cycle, achieving self-sufficiency will take time. Another point to reinforce is the importance of the IT partnership role and, also, why IT cannot be completely cut out of the picture.

Keep focused

Developing reports for the sake of developing reports is inefficient and a waste of resources. The focus should be on developing reports that matter—those that can help in analysis and decision-making. Therefore, focus on the important metrics and key performance indicators (KPIs) which can be pushed to dashboards in an easy to digest format. The recommendation is to provide less but more useful and actionable information. When a platform is easy to use and understand, its popularity and adoption will increase.

Provide the right tool

SSA users interact with the platform through a front-end tool. Its ease of use, navigation, desirable features, and data visualization capabilities will significantly influence how the casual users, who make up the bulk of the platform users, perceive and use it. The selected tool should meet the reporting and analyses needs of the platform's diverse user population.

Market internally

When successes are achieved due to self-service, they should be shared within the organization. The focus should be on highlighting the platform's specific benefits such as cost reduction, productivity improvement, analyses that could not be performed in the past, etc. This can be done through open houses, lunch and learn sessions, tool user groups, tech forums, dedicated SSA website, training offerings, and webinars. Sharing the programs value and benefits will build confidence and bring more users into the platform.

Provide website for the users

The SSA program should maintain its own website, which should be the first stop for users interested in the program. It can provide useful information and also help those who are already on the platform. Such a central site can provide information regarding:

- Program information
- Processes
- Tools
- Onboarding (users and data)
- Roles and responsibilities
- Training materials including step-by-step instructions, best practices, cheat sheets, etc.
- FAQs (pertaining to the program and the front-end tool)
- Forms (onboarding, access request, license request, etc.)
- Governance

- How to get help
- Community information (User Groups—internal and external)
- Links to resources
- Links to reports and dashboards that have been developed through the program

Conduct Open Houses

A series of Open Houses can be used to introduce the program to the masses. Such events can be used to provide information about the program and its status, demo new tools and functionality, answer questions, address concerns, and provide an opportunity where the general user population can interact with IT and the business executives sponsoring the project.

Organize user groups

An SSA deployment strategy should include the formation of a user group, such as the Tableau User Group, which primarily functions as a peer-based support forum. Such a group can provide many benefits including spreading the word about the program, demonstrating the front-end tool's capabilities, enabling users to demonstrate what they have developed on the platform, sharing problems solved, exchanging ideas, and providing tips, as well as demonstrating specific features and functionality that may benefit other users.

Share administration responsibilities

In the typical SSA structure, the tool administrator reports to IT. This is a very technical role which has critical responsibilities and, therefore, IT maintains control over this function. However, a business user (developer) with excellent technical skills who also understands the business can be provided administrator rights. Such an administrator can be authorized to create folders, setup users and roles, and implement permissions for users, reports, and dashboards.

In some cases, developers may be granted permission to migrate objects between environments, such as from QA to production. The benefit of this arrangement is that it shortens the development cycle, reduces the IT administrator's workload, makes the business feel more independent, and creates trust and rapport between the business and IT.

Avoid rigid data governance

A self-service analytics program's success is, to a large degree, dependent on the governance model. It should balance data security, confidentiality, and scalability needs with the flexibility that business users want. If the governance model is rigid and becomes a hurdle for the users, they will stay away from the platform. On the other hand, lack of governance will lead to data chaos, which will cause the users to lose trust in the data and its output. It will also open the enterprise to severe consequences if there are privacy breaches or compliance and regulation violations.

Enable collaboration

Unless there is collaboration between SSA users, silos will develop which will prevent the efficient use of the platform. Therefore, after the SSA program has been rolled out, effort should be made to bring together the various departments and groups using the platform, which will benefit everyone.

An example is the collaboration between sales, marketing, and the customer service teams. Customer service data can be a good indicator of customer satisfaction, which is a leading indicator. If the satisfaction indicator starts trending down, the sales and marketing teams, based on the customer data, can proactively start to engage the customer and take mitigating steps.

The business also needs to have close collaboration with IT, which is responsible for the infrastructure and data. Unless the business collaborates with IT, the platform will face many issues, which will lead to dissatisfied users.

Provide training

When reports are developed on a massive scale in an SSA environment, the potential for incorrect or conflicting reports increases significantly.

This can cause the loss of trust in most reports and, in such a case, users tend to trust their own reports. To avoid this situation, users need to be well-trained, especially in best practices and how to leverage existing content. Unless this is done, the platform will not benefit the organization as it should.

Leverage platform vendor resources

It is better to avoid reinventing the wheel and, instead, utilize what already exists. The analytics platform vendors can be approached to provide documentation and, also, share information about public resources where useful content is available. Typically, these vendors can provide videos, webinars, an online community, and access to some level of free support. In some cases, they can also provide end-user training. These resources should be promoted within the organization.

Chapter 41
SSA LIMITATIONS AND FUTURE

SSA will not replace enterprise BI

Self-service analytics provides users the capability to perform analytics. However, it does not replace enterprise business intelligence, which is implemented and maintained by central IT. Enterprise BI has typically been centralized around data warehouses and data marts, which are accessed and analyzed by heavy-duty BI tools like Cognos, OBIEE, and SAP Business Objects. These data structures and tools, as well as newer tools like Hadoop, can only be handled by very technical and experienced users who are usually part of IT—not by the end-users who form the vast majority of SSA platform users.

In enterprise business intelligence, IT prepares all of the data and also develops most of the reports and dashboards, while the end-users just consume the content. In an SSA environment, IT still retains control and manages the data, while the business users develop most of the reports and dashboards. In this environment, a few users can prepare their own data or bring in new data from outside the platform.

Most of the reports and dashboards developed by SSA users are at the department level. However, enterprise level report and dashboard development continue to be performed under the enterprise business intelligence umbrella. Therefore, SSA is an auxiliary that complements enterprise BI—not replace it.

Analytics professionals will still be needed

SSA provides its users the ability to take care of their relatively simple reporting and analysis needs. However, most users cannot perform complex analysis and, therefore, need the help of more experienced analytics professionals. This help extends to connecting with data sources, developing complex reports and dashboards, collaboration, and publishing objects, as well as interpreting results. Therefore, even though business users want freedom from technical experts, they will continue to rely on analytics professionals for their data, reporting, and analysis needs.

Analytics output of SSA users will grow

SSA has been embraced by all types of organizations in order to democratize the use of data in business. Rapid technological advancements are empowering non-specialists and enabling them to analyze their businesses easily and effectively, which enables better decision-making.

Self-service analytics, despite its challenges and risks, is here to stay. It is considered to be a key strategic technology area by leading CIOs, which has led organizations to implement SSA for creating a data-driven culture. SSA will continue to spread to more organizations as well as within the business hierarchy, from the top levels to the users in the trenches. Therefore, it is no surprise that Gartner has predicted that the analytics output of business users with self-service capabilities will soon surpass that of data scientists.

Collaboration will increase

SSA has liberated platform users who have benefitted from their ability to work independently. For most users, this may be sufficient as it addresses and removes their previous handicap. However, to become more efficient and productive, they need to collaborate with other users within the organization as well as IT. Therefore, it is expected that there will be more focus on collaboration, so that it is enabled seamlessly though tools and workflows.

Tools will become more powerful and intuitive

It is expected that tools which make analysis easier and more intuitive for business users will be the focus of tool vendors. These tools will incorporate more cutting-edge technologies, such as machine learning, predictive analytics, and artificial intelligence. Though human insight is important and essential, there are numerous use cases where machines are superior in their ability to analyze data and provide insight that can lead to better and faster decision-making. It is expected that these tools will be able to automate analyses and provide actionable insights. The embedding of these cutting-edge tools with front-end tools has the potential to deliver in-context automation, analysis, and timely recommended actions.

Embedded analytics demand will increase

Embedded analytics integrates analytic content and capabilities within an application. Since it makes data analysis easier and more accessible to the application users, they can work more efficiently. There will be more demand for this capability in the future and, therefore, it is expected that vendors will respond by developing more embedded responsive tools.

Chapter 42
COMMON PROBLEMS AND ISSUES

Planning ignores user capabilities

The background and skills of business users varies tremendously. Therefore, their platform usage will depend on their technical and business skills, as well as their authorization level, which can restrict their access and ability to execute certain functions. When planning their training, the users' background and skill levels should be considered, as it directly influences the curriculum and content to be delivered.

If user limitations are not recognized, it can create many issues. Users might be assigned the wrong roles and responsibilities, provided wrong permissions, scheduled for the wrong training, or not provided any training at all. Any of these issues can lead to frustrated users, make them non-productive, and increase the risk for the platform.

Self-service limitations

Providing self-service does not mean that a user can do everything in an SSA platform. If we analyze the concept of pumping self-service gas, it is obvious that most drivers can easily pump their own gas. However, that does not mean that they can repair their car using self-service. As is obvious, at some point, an expert is required to provide the appropriate service. The same applies to self-service analytics. Users may need to perform exploration and analysis tasks, some of which are easy while others are complex and difficult. Therefore, they will need help with complex tasks like creating or importing datasets, incorporating complex calculations in a report, or developing a complex dashboard. Novices will need help even with very simple tasks.

Self-service analytics seems to indicate incorrectly, to some inexperienced program planners, that users can perform analytics by simply using a software tool or application. However, the reality is that users will need training, access, and support based on their individual skill levels and requirements. When this is not recognized, the users will not be as productive as expected.

Access provided to untrained users

If users are authorized to access the SSA platform without adequate training and data knowledge, they can create many problems. Every user needs to be adequately prepared before platform access is provided. For example, a data analyst who will need to add or merge data sources will need to understand data and data modeling. If a user does not have these skills and imports new datasets, the potential is high for data issues to develop.

Faulty analysis

When a user working with a state-of-the-art BI tool blindly relies on the results that it generates, it can lead to a decision-making risk. Even though modern front-end tools are easy to use for simple reporting and analysis, even for novices, they need to be used carefully for complex features, calculations, and scenarios. For example, an analysis may require a series of steps which, if they are not executed in the right sequence, can return inaccurate results without the user realizing it. Therefore, to reduce this type of risk, results need to be validated by other users and more experienced analysts.

Just providing tools to users is not enough to make them productive and avoid serious mistakes. Without adequate knowledge and training, the analysis performed by inexperienced users has a fair chance of being incorrect. They can misunderstand or misinterpret data, miss key insights and, therefore, generate flawed results. Such mistakes may not be obvious right away, as their impact may take some time to become apparent.

Issues flowing downstream

During report execution, if an end-user makes a mistake, such as using an incorrect filter or selection period, it will cause an issue which will be limited to that specific report. However, if a mistake is made when preparing data, it will flow downstream into multiple reports and dashboards. Take the example of a data preparation tool used to merge datasets or modify some calculations. If any merge or calculation mistake is made, every report and dashboard using that dataset or calculation will be impacted and produce the wrong results, which can impact thousands of users in a large organization.

If data being used is not clean, consistent, reliable and accurate, the results will also be inconsistent, error-prone, and unacceptable to others using a different dataset which produces a different result. The use of bad data can, to a large extent, be prevented by having sound data validation and governance processes in place.

Inadequate training

Data is an asset that is useless unless it is used for a purpose. Due to its complexity, it should not be handled without training, which can be minimal or extensive depending on the objective and user's role and responsibilities. The risk is that if the training provided is minimal, inadequate or irrelevant, it can have serious consequences such as data mismanagement, incorrect analysis, and faulty decision-making. Users should never be provided access to sensitive datasets or tools without appropriate training. If such training is skipped or inadequate, it creates an avoidable risk.

User onboarding

Departments and groups have different data, reporting, and analysis requirements. The abilities of their developers, power-users, and casual users vary significantly. When such departments, users and shadow IT groups are brought into the SSA platform, they will need help navigating the onboarding process, which can be very simple for some and complex for others. If the onboarding process does not exist or is not robust, and lacks scalability, it will lead to dissatisfied users.

Data onboarding

In addition to the users, data also needs to be onboarded onto the platform using well-defined processes. This data can be sourced from within the enterprise or from external sources. The internal data can be pulled from the managed IT landscape or from a local data source used by a department. Any data from any of these sources must follow an onboarding process, which is in conformity with the data governance rules and guidelines. If a data onboarding process does not exist or is not robust, it has the potential to cause major issues.

Inadequate controls

The users' ability to quickly develop reports and dashboards and then move them into production, where they become available to all authorized users, is a benefit that the business has always wanted. However, if they are inaccurate, they can create many issues. The same issue applies to data. If a user imports a dataset with questionable quality into the SSA platform, it will create issues for all reports and dashboards that use that dataset.

To avoid these issues, there is a need to build controls that can ensure the integrity of the platform. The challenge is to provide a level of control that provides flexibility to the users without compromising integrity, security, and performance.

Lack of governance

A weakness of many self-service analytics systems is governance, which many organizations lack. Lack of robust governance can create many data quality, data security, and privacy issues. Besides the chaos and potential financial loss, it can also lead to the violation of regulations and compliance laws.

In an ungoverned SSA platform, the most common issues expected are the development of data silos, data chaos, data security issues, confusing and unusable data models, incorrect or inconsistent reports, poor decision-making, and widespread duplication of metrics, reports, and dashboards. Other serious issues include loss of single version of the truth, lack of credibility due to incorrect reports and analysis, flawed business logic, and incorrect metrics.

Without governance, compliance, regulatory, and legal requirements will not be satisfied. The organization will become vulnerable to data breaches and reputation damage, as well as significant financial penalties. Internally, it can lead to many issues including lack of trust in the data and results, proliferation of silos, flood of inconsistent and redundant reports, duplication of effort, increased IT support and, after some time, complete chaos.

Shared responsibilities not well-defined

A self-service analytics platform can only be successful if there is a partnership between IT and the business. Therefore, their roles and responsibilities should be well-defined to ensure that they perform their assigned tasks. Usually, IT is responsible for the infrastructure and data, while the business is responsible for developing reports and dashboards. When there is a failure to define or balance the work that needs to be performed by IT and the business, the SSA platform moves towards inefficiency or failure.

Underuse

If self-service tools are difficult to use or too complicated, users will avoid them except for absolutely essential and mandatory tasks. Users need training and hand-holding before they can become comfortable and proficient with such tools. If they are unable to use the tools, they will pass their reporting and analysis requirements to other analysts or IT. Even if users are able to use a self-service tool for basic tasks, it takes significant time and effort to master the tool and produce analyses that can address important business questions and enable effective decision-making.

Another reason for underuse can be the difficulty in preparing datasets and extracts. For experienced developers, these are relatively routine and simple tasks. However, for a business user with less technical skills, this task can be a challenge that can become a barrier. Such users will be forced to get IT help, which will delay the work that they need to perform.

Excessive development

When thousands of users are able to develop reports with limited constraints, thousands of reports can be developed within a short period. This can create issues like duplication, inconsistent reports, distrust in content developed by other departments, lack of standardization, data silos, and data chaos. Excessive and unnecessary development, along with underuse, is a challenge that SSA administrators have to deal with and address, so that there is balance between overuse and underuse.

Misplaced expectations

Vendors have become very adept at convincing customers that their product is easy to use and the most suitable one for their analytics needs. A common selling point is that their tool can enable users to perform awesome analytics through simple point and click operations, which has led to misplaced user expectations. The reality is different, as users have varying skills levels and, for many, their interaction with the tools can be a challenge unless they have been provided extensive and focused training. This risk, with its potential issues, can be avoided through a proof of concept project using real data, which can validate the user-friendliness of the tool as well as its capabilities.

Redundancy

Challenges emerge with the introduction of new technology and processes. An issue that can develop in an SSA environment is that multiple users may be recreating an existing report, dataset, or data model without being aware of it, due to the existence of silos, lack of communication, or poor governance. This leads to redundancy, which can be avoided to a large degree through improved collaboration.

Chapter 43

CHALLENGES

Self-service analytics is an efficient way to enable a data-driven organization and, hence, is expected to have smooth sailing in implementation and adoption. However, its adoption has faced speed bumps for a variety of reasons. In some cases, implementing SSA leads to the data environment getting worse, rather than better. Also, business users continue to depend on IT because they need more than expected support. The following sections describe the challenges faced by SSA.

Uniqueness

Self-service analytics cannot be implemented in an identical way across all organizations, departments, and groups because they do not operate in the same way. Business models and cultures are different and, therefore, they operate uniquely as do users, which means that there is no such thing as a uniformly deployed solution.

There is no one-size-fits-all solution and every implementation is unique. Therefore, any implementation requires careful evaluation to determine the optimal way and schedule for implementing the SSA platform. This should include an evaluation of the business, industry, corporate culture, environment, data sources, role of external data, existing infrastructure, technical expertise of users, analyses requirements, and shortcomings, as well as governance, security, and compliance requirements. Therefore, any SSA program that skips an in-depth evaluation and uses generic implementation templates will fail.

Organizational dynamics

In organizations, conflict between the bottom-up and top-down approaches is fairly common. In the conventional business intelligence environment, the top-down approach is pushed by senior management because they desire centralization, standardization, controls, business alignment, economies of scale, and availability of a system that provides a single source of truth.

In self-service analytics, the top-down approach conflicts with what the rank and file users desire. They want autonomy so that they can meet their department's needs flexibly, quickly, efficiently, at minimum cost, and without being constrained by formal processes and checks that slow the development process.

To ensure success, the differing goals for top-down and bottom-up approaches need to be recognized and balanced when implementing a self-service analytics platform. Failure to provide this balance will lay the foundation of failure.

Enterprise versus local conflict

The inability to reconcile differences between the enterprise and local needs for standards and controls can create issues for the platform. If platform access is too restrictive, it will turn away users. On the other hand, if it is too open, it can lead to data security and privacy issues, which can have a devastating impact when there are breaches of any kind, internal or external.

To avoid this issue, the data governance policy and rules must be agreed to and approved by both sides, which will avoid resistance and ensure adherence. It will also ensure that flexibility and speed are provided without putting data at risk. When data can be accessed easily and in a safe manner, it satisfies a precondition for a successful self-service analytics implementation.

Loss of control over extracts

When there are too many requests for producing extracts for reports and dashboards, it can create many issues. In many cases, the demand for a new extract is unnecessary because an existing extract can be used. Such a request can frequently be attributed to the requestor being unaware of its existence. These requests create an extra unnecessary load on the limited IT resources supporting the SSA platform, increase redundancy, and potentially impact the ETL process.

Cultural barriers

An SSA platform, contrary to the perception of many users, is more than just a tool for developing reports and dashboards. To make its users view SSA as more than a reporting tool that pushes content to dashboards, organizational and cultural barriers need to be overcome. The platform needs to communicate and collaborate with its internal users, as well as customers, partners, and vendors. For organizations that can overcome this challenge, the platform will be successful.

The transformation to a data-driven culture is usually slow. It requires that IT and the business develop a partnership and a commonly agreed strategy, which should include a change management component. Together, they can help change the culture and determine the best way for optimizing the use of the available tools and resources.

Balancing agility with access controls

SSA users, typically, are in a hurry and want access as soon as a reporting or analytics need is identified. However, IT is responsible for supporting the tools and maintaining a curated system whose access is subjected to strict control. Therefore, users have to work within these constraints, which they view with frustration as it causes development and rollout delays.

Controls and governance

Some users are disappointed when they are unable to accomplish what they want to do, often due to misplaced expectations. Since an SSA is a governed and secure environment, access is provided after many prerequisites have been met. These can include tools training, governance training, authorization for specific datasets, user role setup, platform access authorization, authorization to work with PHI and PII data, etc.

Each prerequisite requirement can lead to the perception that the platform is too bureaucratic and cumbersome. Additionally, after users start using the system, they may run into other control issues. For example, users cannot promote the objects that they develop in the development environment to the production environment. This is a task usually handled by IT, which is responsible for maintaining the platform infrastructure, setting up folders for departments and users, authorizations and permissions, security, performance monitoring, etc. This issue can be avoided by educating the users through training and managing their expectations.

Regulation and compliance

With privacy breaches becoming a norm, organizations are very sensitive to how their data is used and protected. They are also required to follow regulations like the European Union's General Data Protection Regulation (GDPR), Health Insurance Portability and Accountability Act (HIPPA), etc. Any breaches or compliance failures can have significant financial and reputational repercussions.

As part of the effort to meet regulatory, compliance, and privacy requirements, organizations need to track data lineage, which is the data lifecycle. Data lineage starts with the data's origins and covers its movements, characteristics, and quality over time. It describes what happens to data as it goes through diverse processes, capturing all the steps and changes from the time data is captured to the current time. Implementing these requirements can create hurdles for those trying to use the data. This is a challenge that must be addressed by all types of implementations.

Shortage of skilled data professionals

An SSA platform can have thousands of active users, who can provide useful analyses. However, enterprises also need experienced data analysts and data scientists, who can provide key insights that ordinary users cannot. These highly skilled professionals can perform in-depth analyses from different perspectives because they are very knowledgeable about the business and its data. However, when such a platform has less than the desired number of experienced analysts, it increases the risk of the platform performing below its potential.

Chapter 44

REPORTING AND DATA CHALLENGES

Limitations of traditional tools

The most widely used tool for analysis is Excel. However, it was not designed for large datasets, scalability, governance, advanced charting, data manipulation, or collaboration. Traditional BI tools have limited capabilities and are not user-friendly, which limits the reporting that users can perform. Therefore, a key user requirement is for a front-end tool that enables easy reporting and analysis. If such a tool is not provided to the users, they will become frustrated due to their inability to perform reporting and analyses quickly, easily, and safely.

Another issue is that the most popular self-service analytics tools used by business users are designed to run on small datasets on desktops, laptops, or the web. However, in many cases, when users try to analyse large datasets, the performance of these tools is poor.

Decentralized reporting

In a self-service environment, having a large number of users and departments develop reports and dashboards can inadvertently lead to reports that are based on data of varying quality and, also, provide inconsistent and faulty results. These problems multiply when such reports and results are shared, leading to widespread flawed decision-making.

Multiplication of redundant reports

Business users want easy-to-use tools that can help them analyze data and produce reports quickly without help from IT or analysts. Since many of these tools are inexpensive, individual departments buy them independently, which leads to the setup of shadow reporting systems. When users are provided front-end tools that they previously did not have access to, they quickly start churning out reports and dashboards.

In such a scenario, there is usually no communication across departments and reports are developed in silos. When development occurs in such an unplanned manner by a large number of users, it leads to unchecked multiplication of redundant reports. This creates chaos and many issues such as inconsistent and inaccurate outputs across business units, reports that are not trusted except by the report developer, and inaccurate reports are produced even when IT managed data is used. It becomes difficult to find the correct report, due to the explosion in the number of reports, and one version of the truth becomes a casualty.

Inconsistent results

Without IT-managed curated data, results generated can be confusing because the "same" reports produce different results. Metrics and KPIs are not calculated using a standard method and can be based on inaccurate data. When an analyst quits, addressing report issues or enhancements will be a challenge, due to lack of available documented information. In such a case, IT may need to get involved in resolving inconsistencies because it manages the datasets. To avoid such a situation, policies need to be in place to ensure that reference documentation is developed, as part of the governance process.

Excessive focus on visualizations

Many self-service analytics users tend to focus on developing dazzling visualizations, rather than developing content that can provide valuable insight into the business. Visualizing data is a very powerful way of presenting results. However, it can become an issue when it causes data exploration and querying to take a back seat, which prevents users from obtaining actionable insights.

Data silos

If departments create their own data models and data structures (data marts) in an SSA environment without IT involvement, they can perform their own analyses and achieve independence which they had lacked previously. In a scenario where departments work independently, without leveraging common data sources or adequate governance, many data silo issues develop including inefficient data models and inconsistent metrics. Therefore, when reports based on such data models or metrics are shared with other departments, poor performance or conflicting results can be expected.

Data quality

Data quality is a challenge wherever data is captured, stored, and analyzed. The degree of success achieved in maintaining data quality varies across organizations because their approach varies significantly. The most successful ones implement a sound data governance strategy that is well documented, communicated, and enforced. Organizations that do not have data governance in place or where it is not enforced, frequently have to deal with poor data quality issues, which can negatively impact report results and lead to faulty decision-making.

Data complexity

Data is growing exponentially, more complex and, also, no longer limited to simple relational database sources. Data sources now include structured and unstructured data. The data types being used include clickstream data, video, voice, social media, real-time, etc. There is increasing use of Big Data, which is defined by volume, velocity, and variety characteristics. Therefore, data management can become a very complex process, especially when large volumes of data are imported from multiple sources in different formats. Such data needs to be linked, matched, cleansed, and transformed across systems. This is a challenge for SSA environments and cannot be ignored without negative consequences.

Data preparation

Data preparation is one of the most complex, difficult, and time-consuming challenges that business intelligence users face in legacy BI environments as well as SSA platforms. In a traditional BI environment, IT has complete responsibility for this task. However, the SSA analytics environment is complex due to shared responsibilities. While IT prepares most of the data, self-service users also need to prepare some data. This can be an issue because the technical capability of SSA users can vary quite a bit and, therefore, many users may need IT help even for relatively simple data preparation tasks. While IT will provide the help needed, this creates two issues. The first issue is that there will be a delay, which SSA was meant to eliminate. The second issue is that such users will not be as independent as they desire in an SSA environment.

New data sources not prepared correctly

Data drives everything in an SSA environment. The effectiveness and efficiency of this critical element directly impacts the quality of analyses and decision-making. Data is served through processes which need to be automated as much as possible to reduce errors. The issue is that as new data sources are brought in, automation may not be immediately available. Also, there is a fair chance that analysts preparing the data may not be experienced enough to ensure that it has been prepared the right way. This can be avoided by involving skilled IT resources for technical validation, along with business users who can validate it from a different perspective.

Downside of data democratization

Data democratization is very beneficial but can create many issues if there is inadequate governance. While freedom to work with and explore data is to be encouraged, there needs to be a check on the access provided for critical business data. In SSA environments, even though it is challenging and resisted by some key users, access authorizations, governance rules, procedures, and policies should be carefully implemented in a balanced way so that risks are minimized, while providing maximum flexibility to the users.

JSON challenge

JavaScript Object Notation (JSON) is a lightweight, open-standard data-interchange file format for storing and transporting data, which is easy for humans to read and write. It is also easy for machines to parse and generate. The issue is that the tools designed for SQL do not handle JSON well and, also, blending JSON data with other data is a challenge. To deal with JSON, the services of IT are required, which is a step back for self-service.

Chapter 45

CONCERNS

Business users' concerns

If business users create data silos, intentionally or unintentionally, they end up outside the enterprise data architecture. In such cases, the business will end up supporting their own data sources, extracts, reports and dashboards, which is an undesirable outcome as it can be costly, inefficient, resource intensive, time consuming, and difficult. It will also lead to duplicate processes and redundancy. From an enterprise perspective, this is unnecessary and, also, not cost effective due to the required duplicate effort.

The existence of silos also means that users have a narrow view of the business and, therefore, they cannot conduct cross-functional or cross-departmental analysis. Such silos produce reports on KPIs and metrics that are not governance approved, which can lead others to question their validity and accuracy.

Enterprise strategic concerns

SSA users can get up to speed quickly and become very productive. However, while this is welcome from their own business unit or department's perspective, it can be a negative from the enterprise perspective if they work in silos with data that is not shared. The concern is that data structures built and maintained in silos are often developed without any strategic enterprise considerations.

IT concerns

For IT, which is responsible for maintaining the infrastructure and data integrity, the development of silos and numerous department data stores is a serious cause for concern, due to the potential issues described previously. In such an environment, it is not easy to validate data, reports, or dashboards. Any validation effort is slow, difficult and resource intensive. It is even more challenging when data is sourced externally.

Another IT concern is that inadequately trained users may be authorized to work in the system. Such users can develop incorrect reports and dashboards, create more help desk tickets, and need more support for issues pertaining to the back-end as well as the front-end. This issue can be avoided by making relevant training mandatory.

Analysis delay

Data being collected and stored is growing exponentially. The problem is that most organizations and users are barely scratching the surface of the available data. While data collected over a period of time can be used for analysis, such as trend analysis, it loses its value if it is not used with a reasonable time. For example, analysis of ski sales is beneficial in the spring because the previous winter's data can be used to increase or decrease next winter's ski production. However, if the analysis is delayed till the fall, it will be of no value because there will be insufficient time to decrease or increase the production for meeting next winter's demand.

The challenge is to quickly identify the most promising data and then analyze it. This can be easier to achieve in an organization that has a data-driven culture. If such a culture is lacking, a strategic plan should be developed to bring about the required change.

Additional O&M costs

Self-service analytics is eagerly welcomed by the business due to the benefits that it can provide, especially self-sufficiency. However, there is also a concern due to the additional costs that are associated with its implementation as well as ongoing operation and maintenance. These include costs associated with licensing, security, governance, and infrastructure, which will need to be managed by highly skilled IT resources.

User limitations

As the available enterprise data grows exponentially, the ability to provide accurate insight is limited for users with limited technical or business knowledge. A complex front-end tool can become a dangerous weapon in the hands of users who do not know how to use it correctly. Also, in many cases, when too much is data available, users simply do not know where to start unless they have been provided relevant training.

How to keep SSA relevant

The rising popularity of SSA, along with the recent development of data preparation and data visualization tools, has raised questions about the long-term viability of data warehousing. Data warehouses and data marts have been the backbone of data analysis and reporting for the past quarter century or so. However, there has been a paradigm shift that has reduced its importance. This is how technology evolves, which leads to its partial if not complete replacement by another technology.

We can expect the same for self-service analytics, which will undergo enhancements till another technology will replace it or make it less important. The challenge is to keep SSA relevant for as long as possible by proactively adding capabilities that it currently lacks. Its key challenge includes incorporating technologies like artificial intelligence (AI), machine learning, and predictive analytics, which have the potential to make analytics even more powerful and easier.

Chapter 46

BAD PRACTICES AND MISTAKES

SSA programs can provide many benefits if they are implemented correctly. Despite successes, many projects have not realized their potential benefits. Many of the mistakes leading to shortcomings can be attributed to practices that should have been avoided. These mistakes are made at different stages including planning, design, implementation, tool selection, training, and rollout. The following sections describe common mistakes, along with pointers on how to avoid them.

Using poor quality data

When data of poor quality is used, it becomes the foundation of faulty decision-making. Data quality issues are pervasive in organizations, which leads to unreliable and inaccurate data and analyses. Basing analyses and decisions on such data can be very costly. Fixing data issues takes time and resources, which means that other important tasks take a back seat. Fixing issues is expensive, causes frustration, and makes users lose trust.

A better way is to fix issues before they create any problems. This can be done by implementing data quality management and master data management, which will help prevent or correct bad data before it is served to the users.

Underestimating data preparation effort

Preparing data may require many operations. The process starts with profiling, which involves surveying the data sources in order to understand their condition. This is followed by cleansing, matching, and merging before the data becomes suitable for consumption from accuracy, completeness, timeliness, and consistency perspectives.

These tasks are very resource-intensive, though the introduction of data preparation tools has simplified them. A common mistake is to underestimate the effort required to prepare data, with or without such tools. A common reason for that is the belief that the available source data is clean or not very dirty—an assumption that frequently turns out to be incorrect. A better approach is to assume that the data is dirty and will need to be cleansed before it can be used.

Failure to integrate SSA with BI strategy

If a self-service analytics program is not used as part of the enterprise business intelligence strategy, it will not be as effective as it should be. In many organizations, SSA tools are rolled out to the users who start developing content in silos without any enterprise considerations. While this serves the short-term needs of those users, it limits the effectiveness of the platform. An implementation that addresses the strategic needs, while empowering users, will provide more benefits and deliver the best value to the business. Ideally, a single comprehensive platform, with capabilities and tools that can meet the needs of a diverse group of users, should be implemented instead of a platform with a narrow focus.

Ignoring non-traditional data sources

The traditional BI data sources have been data warehouses, ERP, and CRM systems, which provide very valuable data for analysis. However, in the Internet age, these sources are not sufficient as there is a gold mine of data available through other sources, both structured and unstructured. Examples include social media (Facebook, Twitter), smartphones, e-mails, YouTube, clickstream data, machine generated data, IoT, etc.

These diverse sources can provide data, much of it in real-time, that can be very valuable. For example, an organization used data from multiple sources to calculate the current customer sentiment index. The sources included an OLTP database, Help Desk tickets, audio messages, e-mails, and Twitter feeds. By analyzing the data using machine learning, the company was able to forecast the customer sentiment index, which is a leading indicator. The analysis helped identify customers whose sentiment index was decreasing. Since such customers pose a high risk of churning, they were contacted proactively to address their concerns, so that they could be retained as customers.

Standalone implementation

For SSA to be successful and widely adopted, it should not be implemented as a standalone platform. The most effective application of analytics occurs when users are able to access the platform from within their operations application and workflow. An example is the ability of a user to access an enterprise dashboard from a smartphone, which gives that user the freedom to perform analysis anytime and anywhere.

Another example is the ability of a user to use an analytics tool to drill-down to the underlying data, located in the OLTP database feeding the business application. For example, after viewing a dashboard report, the user can navigate to the underlying data stored in the database through a few simple clicks.

Incorrect use of visualizations

Data visualization is the technique for presenting data in a graphical format. Its primary goal is to communicate information clearly and efficiently to users using information graphics, such as tables and charts. Effective visualization helps users analyze data by making complex data more accessible, understandable, and usable.

A common mistake is to develop visualizations based on their aesthetic appeal, rather than the character of the data. The key to using data visualization effectively is to use the most appropriate type of visualization. For example, tables should generally be used when there is a need to display a specific measure of a variable, while charts of various types should be used to show patterns or relationships in the data for one or more variables. Bar charts are used for comparing data across categories, while line charts are used for viewing data trends. Other common mistakes include overuse of pie charts and color.

Lack of standards

An environment used by thousands of users will contain thousands of reports and dashboards. If such an environment has no standards, it will be characterized by inconsistency. For example, naming conventions as well as the look and feel will be different. This can be confusing for users who want to leverage what has already been developed by other departments or groups. Implementing standards, such as UX standards, can help avoid some of these issues. Standardization can help users interpret datasets and metrics quickly and, also, improve collaboration.

No data governance

Data governance has many benefits and must be an integral part of any SSA platform. Having no data governance is one of the worst mistakes that can be made in such an environment. While the SSA platform empowers users and provides them flexibility, which is essential for the success of the platform, it cannot be done by compromising the safety of the platform and the organization.

Governance must be implemented in a systematic manner and not as an afterthought. A common mistake is that data governance is approached casually, which must be avoided. Implementing a well thought out governance policy will benefit users as well as the enterprise, despite a controlled and structured environment.

Chapter 47

BEST PRACTICES

The implementation of a self-service analytics project empowers users and provides many benefits. However, making the transition from traditional business intelligence to self-service analytics is challenging and not smooth in many cases. To meet these challenges, a few best practices and rules will help develop a solid foundation that will enable organizations to leverage and scale SSA capability. These best practices and tips are described in this chapter.

Build a robust data foundation

Having a robust data foundation for an analytics program is absolutely essential. It should serve reliable, clean, and trusted data which can be accessed securely and easily. The foundation can be a data lake, data warehouse, or a multi-platform system. Due to its critical role and the technical skills required to operate and maintain it, IT is responsible for this infrastructure.

Architecture should support self-service

Providing the users with a front-end tool which is not integrated with the platform architecture is a mistake, which will ultimately lead to serious issues. Organizations must first develop an architecture that will support the self-service platform. It should be architected so that it clearly maps the users with the infrastructure components and technology.

The architecture should clearly define and support the process for migrating data, reports and dashboards across environments, from development to production via quality assurance. This should be complemented by an organizational matrixed model that balances the centralized and decentralized forces, which come into play when the platform users and the IT/governance teams try to go in different directions.

Maintain data quality

Data quality is an indicator of the effectiveness and reliability of data. If data fails to meet data quality standards, the platform will be shunned by the users and fail to meet its objective. When data is clean and trusted, the analysis produced will also be trusted. If the data is inaccurate and of poor quality, it will lead to a loss of confidence in the results that it produces. Bad data is also costly because of the time and effort required to rectify the issues caused by it.

There are scores of attributes that have been used to define quality data. However, the characteristics frequently associated with good quality data are accuracy, completeness, conformity, consistency, integrity, timeliness, availability, auditability, and relevance. Data of high quality is understood and trusted by users, lacks duplication, and adheres to data standards. These characteristics must be maintained throughout the data lifecycle.

Develop standards

On a platform with thousands of users and scores of groups, chaos will reign unless there are standards and controls in place. Developing and enforcing standards has many benefits. Metrics and data definitions will be consistent and trusted. Duplication of metrics, reports, and dashboards will be avoided. Also, when the look and feel of reports and dashboards is consistent, it will help users to perform analysis quickly.

Develop a certification process

If reports and dashboards are certified for accuracy based on an alignment with the company standards, it will build confidence in them. This will lead other teams to seriously consider using them, which will help avoid duplication, reduce the overall development cost, and lead to trust in the analytics platform.

Develop security standards

Securing an SSA platform cannot be a reactive exercise. Security standards and policies should be developed as early as possible. Data is a corporate asset and should be secured before access is provided to anyone. The level of security should be such that the platform and data are protected, without making access so cumbersome that it turns away the users.

Use data preparation tools

Getting data ready for analysis is one of the most difficult, slow, and tedious tasks in the analytics process, especially for non-technical users. This becomes a real challenge when the data is of poor quality or is located in different data sources. In the past, this task used to be performed by IT. However, with the introduction of data preparation tools, this task can be performed by less-technically savvy users with relative ease. These tools should be leveraged to the maximum as they can reduce the data preparation time and, also, help improve data quality. Providing data prep training will enable more non-technical users to work with and use data, a capability that they did not have in the past.

Provide customized training

Training is often not provided the importance it deserves. It is not sufficient to just provide users with access to an analytics tool and then expect them to perform meaningful analyses. Vendors tend to exaggerate when describing their tools. They often claim that pointing and clicking will make a user productive. The reality is quite different. Every user needs to be trained, though the level and breadth of training will vary, depending on the tool's complexity and the user's skills and role.

The training to be provided should be broader than just the tool training, which is very important. SSA platform users need to be trained, as applicable to a specific role, in the processes, data and data analysis, collaboration, security, and governance.

Leverage power-users

Power-users, who are advanced users placed between casual users and IT, are a valuable resource who should be leveraged as much as possible. They reduce the IT workload and backlog and, also, become the first line of support for department users. They create reports and perform ad hoc analysis for themselves, as well as for their department's casual users. A network of power-users can develop a grassroots foundation that can expand the analytics footprint, while also providing support.

Power-users need to be provided maximum flexibility so that they can be productive to the maximum. However, that needs to be balanced with controls, such as for publishing, so that security and confidentiality are maintained. While power-users should be permitted access to advanced tool features and authorized to share their development, they should work under controls that are reasonable and not restrictive.

Operationalize analytics

Front-end tools enable important facts and insights to be highlighted. However, they do not provide any guidance on how that information can be converted into meaningful action and, consequently, results. In most environments, analysts and decision-makers make decisions in a reactive way after viewing the facts. However, to obtain maximum value, users should be empowered to act, based on the available insight. This can be achieved by embedding analytics directly into the business application and processes. When such information becomes available as part of the user's operational activities, within the routine workflow, it can be used very effectively.

Go Agile

Waterfall used to be the most popular methodology used for BI applications and projects, despite its inflexibility. It had many shortcomings, with the most serious one being that it took a long time to implement. The period between the identification of a need and its implementation, through a project, was often too long. In many cases, the initial requirements changed by the time the project was rolled out.

In the Waterfall methodology, many mistakes cannot be identified till very late in the project and, also, users do not see the results till the very end. When the end product or functionality is rolled out, it may not be what the users want, even though they wanted it months ago. In such a case, any rework can be time consuming and costly. Also, fixing any issue with the product or functionality, delivered as per the original requirement, can be time consuming and costly to fix.

The Waterfall methodology shortcomings led to the popularity of the Agile methodology, which uses an iterative approach to roll out incremental functionality or products. It is especially useful when there is an urgent business requirement that needs to be implemented quickly. In this approach, projects are implemented in multiple sprints, where each sprint leads to the rollout of some feature or product. The benefit for users is that they can see the requirements being developed and rolled out in increments, which they can start using immediately. For example, if 4 dashboards are to be developed in a project, they can be developed in 4 increments over 4 months. This means that the users will receive a new dashboard every month.

For analytics, the Agile methodology along with the self-service approach has resulted in more agile and responsive enterprises. Agile should be the preferred methodology for analytics projects, especially for projects where new data is being brought into the platform.

Chapter 48

LESSONS AND TIPS

Adjust to the new reality

Both the stakeholders in an SSA environment, business and IT, are significantly impacted by its introduction, which replaces a centralized system with a decentralized one. It changes the culture and power structure, which requires adjustment and realignment. IT has to accept that many of its responsibilities, like report and dashboard development, have shifted over to the business. The business has to realize that it has more responsibility in the new environment and cannot dump minor requirements over to IT.

Have realistic expectations

A widespread perception is that self-service analytics empowers all business users when they are provided easy-to-use tools. Some organizations believe that rolling out static reports and dashboards with some interactive ad hoc functionality will meet the analytics needs of all users. This is an incorrect assumption because platform users have different needs and requirements.

Some senior managers may be satisfied with dashboards that can be analyzed at a glance, while others may like to receive real-time alerts when metric thresholds are breached. More advanced users will not be satisfied unless they can perform advanced analytics, source their own datasets, combine datasets, prepare or manipulate data extensively, develop advanced visualizations, and perform other more advanced tasks. Therefore, if management has any unrealistic and misplaced expectations, they should be adjusted so that they are in sync with reality.

Encourage collaboration

IT and the business must develop a collaborative relationship or there will be avoidable inefficiencies, development of silos, and other serious issues. In the most successful SSA implementations, there is close collaboration between the business and IT. Though IT is no longer at the forefront of creating reports and dashboards, it provides a valuable service by preparing data for the users. Also, when needed, it helps power-users by troubleshooting issues, creating new metrics, as well as developing complex reports and dashboards.

When they work together, IT gets a good sense of what the business wants or is moving towards, which helps it become more proactive and responsive to the business needs.

Organizational structure must support the program

There are many factors that determine the success of an SSA platform, with one of the most important being the organizational structure, which must support the program. To ensure success, the SSA program and organizational goals should be aligned and specified in measurable terms.

Develop IT and business partnership

An SSA program is a business program, which requires close cooperation between the central IT team and the business. For success to be achieved, they need to develop a partnership through every phase including requirements, design, development, rollout and support. They need to be clear about their responsibilities and, also, know where each needs to take the lead or provide support. This is a relationship where both sides need each other and, therefore, they must have an excellent working relationship for a positive outcome.

Modify legacy IT organizational structure

The traditional IT/business intelligence organization structure and workflow will not work in the new paradigm. In the SSA platform, users are empowered and their dependence on IT is reduced significantly. In the legacy process, IT receives a report request and puts it in the development queue. After doing that, it will typically have no contact with the users till the product is ready for user testing or rollout. In the SSA environment, IT has to remain in close and constant contact with the business. Its goals and incentives are also aligned with the business and, therefore, there is a mutual desire for achieving their objectives.

Since the roles and responsibilities of the business as well as IT changes significantly, the SSA platform must be supported by a modified IT organizational structure. In the new structure, business developers and power-users need to be included, for routine work as well as for projects, because they also have technical responsibilities.

Implement governance and controls

A self-service analytics platform must have a solid governance program and a control system in place. A typical setup is a committee that includes power-users, business analysts, subject matter experts, as well as members from IT and the governance team. This committee should meet periodically once or twice a month to review the current status, ensure governance rules are being followed, resolve issues, and plan for the future. The monitoring of governance rules should be supplemented with periodic audit checks.

For the platform itself, IT needs to implement controls which will ensure that only authorized users are able to access data, reports, and dashboards. Users should only have access to authorized tool functions and features. Unless authorized, they should also be restricted from sharing or downloading data and other content. Processes for monitoring compliance and ensuring security must be implemented and enforced.

Enable flexibility in governance

Governance can be frustrating for users as it tends to slow them down. While this control mechanism is essential and all its rules must be followed, there are situations when flexibility will not expose the data and platform to risk. Therefore, when a business request can be implemented without any potential issues, flexibility can be applied in applying governance rules. In such a case, besides helping the users, flexibility helps to enhance IT's relationship with the business, who will view IT as partners rather than as a bottleneck.

Never follow a one-size-fits-all approach

Since an SSA platform has many types of users, it is simply not possible to address their needs and requirements with a single approach. The data and analyses requirements of data scientists, developers, tool administrators, power-users, executives, and casual users will be quite different. Their needs will be different pertaining to the processes, data, tools, features and functions (such as filtering and downloading), dashboards, reports, collaboration, device requirements, access requirements, training, documentation, support, etc. Therefore, a one-size-fits-all approach for users must be avoided for an SSA platform.

Evaluate tool from different perspectives

Before a tool is selected for self-service, it should be carefully evaluated from different perspectives, technical and business, to ensure that it is suitable for self-service. Examples of desired capabilities include administration, collaboration, data visualization, support for access through various device types, ability to use different types of data sources, in-memory capability, etc. The tool should be able to support current needs, as well as needs expected to develop over time. Serving a short-term need will turn out to be more expensive and disruptive in the long run.

Provide ramp-up help to users

After users are onboarded onto the platform, they will start developing content, which takes time. With thousands of users, many of them new, a fair amount of time will be spent trying to figure how to use or develop reports, dashboards, and metrics. Therefore, it will help the users if they have access to a starter library of report templates and standardized metrics that they can start using right away. Besides making the users more productive, it will also help in maintaining standards and consistency.

Provide training before and after rollout

Self-service analytics program planners are surprised when they realize that the users need more training than they expected. This is not surprising because most of the tools are not intuitive, even for technically savvy users like the power-users. Users need to be trained because it takes time to learn and become proficient with such tools. The breadth and depth of training required will depend on the user roles, responsibilities, and skill levels.

For a self-service analytics program to succeed, training should be incorporated in the project plan for the initial rollout. Typically, post-go-live, the need for advanced training develops quickly as users get up to speed and want to do more with the available tools. Therefore, there should be a plan to provide continuous training for existing users, as well as training for new platform users.

Develop an onboarding plan

Self-service sounds easy but in reality, getting users started with SSA is not easy or smooth, especially for the first batch of users. Therefore, after the platform processes are defined, a detailed onboarding plan should be developed for the users, as well as for the data which needs to be brought into the platform. This may require the users to be trained and certified, followed by managerial approval for accessing the platform. Access should be based on their specific role, which will determine what they can access and what they can do on the platform.

After the users are approved, they will need to be setup by IT in the SSA environment with specific roles. They will also need to be setup with folders, if they are allowed to save reports and dashboards on the platform. Having a formal onboarding process in which some steps are automated will lead to easier and faster onboarding, which will help avoid user frustration.

Made in the USA
Middletown, DE
03 September 2021

47576296R00119